HUMAN FACTORS CHALLENGES IN EMERGENCY MANAGEMENT

Human Factors Challenges in Emergency Management

Enhancing Individual and Team Performance in Fire and Emergency Services

Edited By
CHRISTINE OWEN
University of Tasmania, Australia

CRC Press
Taylor & Francis Group
Boca Raton London New York

CRC Press is an imprint of the
Taylor & Francis Group, an **informa** business

CRC Press
Taylor & Francis Group
6000 Broken Sound Parkway NW, Suite 300
Boca Raton, FL 33487-2742

First issued in paperback 2017

© 2014 by Christine Owen
CRC Press is an imprint of Taylor & Francis Group, an Informa business

No claim to original U.S. Government works

Version Date: 20160226

ISBN 13: 978-1-4094-5305-5 (hbk)
ISBN 13: 978-1-138-07166-7 (pbk)

Visit the Taylor & Francis Web site at
http://www.taylorandfrancis.com

and the CRC Press Web site at
http://www.crcpress.com

Contents

List of Figures

List of Tables

List of Contributors

Dr Chris Bearman
Chris is a Senior Research Fellow and Program Director for the Masters in Safety Science at the Appleton Institute of Central Queensland University. Chris conducts industry-focused research in the areas of occupational health and safety, human factors and applied cognitive psychology. Chris has worked closely with industry partners and government organisations around the world to produce research that has both a strong theoretical underpinning and a robust application to industry. Some highlights of his research are working with small commercial aviation operators to determine the pressures that lead to sub-optimal flight-related decisions, identifying high risk tasks and mitigating strategies in the volunteer incident commander role, helping rail operators to develop an evidence-based approach to evaluating new technology, and working with the National Aeronautics and Space Administration on projects that sought to redesign the US airspace system. Recently, Chris has investigated breakdowns in coordination at the incident management team and above during emergency management.

Dr Karyn Bosomworth
Karyn is a Post-Doctoral Research Fellow in RMIT University's Centre for Risk & Community Safety, and its Climate Change Adaptation Program (CCAP). With over a decade of conducting policy-relevant research and analysis within government, her academic research interests are policy-orientated and interdisciplinary, including climate change adaptation; adaptive governance; disaster risk management; institutional and frame analyses; and the research-policy interface. She enjoys involvement in a range of collaborations across these arenas.

Peter Bremner
Peter is a part-time PhD candidate at CQU whose research includes pressures on decision making and mitigating strategies used by bushfire incident controllers both on the fireground and at higher incident management levels. Peter is also an experienced systems engineer and a volunteer firefighter.

Dr Benjamin Brooks
Ben is a human factors researcher and Senior Research Fellow in the National Centre for Ports and Shipping at the Australian Maritime College. Ben has been a researcher and safety management system consultant for 15 years. He currently works on research in areas such as advanced training and safety systems and safety information systems, and builds tools for the evaluation of safety management systems and safety culture to support individual and team cognition. He works

with a range of stakeholders, including regulators, private companies, pilotage organisations and port authorities.

Steven Curnin
Steve is a PhD candidate at the University of Tasmania and his interest is investigating how stakeholders in emergency management use boundary spanning activities in temporary supra organisations to facilitate multi-agency coordination.

Dr Jan Douglas
Jan has recently completed her PhD exploring the role of individual and collective affect in fire incident management. Her study took a sociocultural perspective based on the belief that people and their work contexts cannot be separated. Jan has worked in a range of high-reliability industries, including aviation, emergency medicine and fire and emergency services.

Dr Lisa M. Frye
Lisa is a human factors researcher. Her PhD studies explored how people make decisions during large-scale bushfires; in particular, how bushfire fighters and residents regulate their thinking while they are under pressure. Lisa's current role is to develop the leadership and decision making skills of bushfire fighters and incident management teams who respond to emergencies in Victoria.

Dr Peter Hayes
Peter has just completed his PhD. His research used specially created bushfire simulations to investigate the human factors underlying differences in teamwork and decision making between pre-formed and ad hoc incident management teams. Prior to undertaking his studies Peter worked for 15 years in forest management, including fire management.

Dr Claire Johnson
Claire completed her PhD examining the consideration of possible worst case scenarios in fire incident planning. This work has introduced the pre-mortem technique as a method for overcoming planning inertia and has generated a set of wide ranging recommendations for fire and emergency services agencies. She has also developed human factors interviewing and debriefing protocols suitable for use by staff.

Andrew Lawson
Andrew holds the position of Deputy Chief Officer at the Country Fire Service (CFS) of South Australia. He has extensive experience in all areas of fire and emergency services, including operational firefighting, fire planning and mapping, administration, management and the development of policy. Andrew was selected by the Australasian Fire and Emergency Service Authorities Council (AFAC) as a

member of the national steering group that developed the new Australasian Inter-service Incident Management System (AIIMS) 4th edition manual.

Professor Jim McLennan

Jim is an industrial/organisational psychologist. He is interested in safety-related decision making. He has worked with the Melbourne Metropolitan Fire Brigade investigating decision making by on-scene Incident Controllers. This has led to various research studies concerning decision making with the Defence Science and Technology Organisation—Air Operations, and Marine Operations, the Office of Corrections and CFA. Since the 2009 Black Saturday Victorian bushfires he has been working as a member of the Bushfire CRC Task Force undertaking research in the aftermath. His current research activities include the role of anticipation in managing complex tasks, emotional and cognitive self-control in survival-related decision making, and effective community alerting and warning.

Dr Mary Omodei

Mary is a cognitive psychologist whose overall research program focuses on the human factors underlying decision making in complex systems, including military command and control and emergency management. She has developed the Networked Fire Chief simulation software and the helmet-mounted video debriefing technique. Mary led the Bushfire CRC 'Safe Behaviour and Decision Making' and 'Volunteerism' projects.

Dr Christine Owen

Christine's research investigates communication, coordination and collaborative practices in high-technology, high-intensity and high-reliability environments. She has conducted research in aviation, emergency medicine and emergency management environments. She has a particular interest in sociocultural theories of work activity and how learningful and developmental work environments may be enabled.

Dr Sue Stack

Sue is an experienced educator and facilitator in a variety of learning formats, including face to face, group-work and online learning. Sue also has extensive experience in designing online learning communities and curriculum development in fire and emergency services.

Roger Strickland

Roger holds the substantive position of Senior Instructor Wildfire in the Country Fire Authority Victoria (CFA), is a Wildfire Investigator, the CFA State Planned Burn Co-ordinator and a member of the CFA Serious Incident Investigation Team. He has 33 years of fire experience, qualified at level 3 in Operations and Planning, and is actively involved in pioneering techniques in incident debriefing, investigation and wildfire training incorporating current human factors research with a focus on improved learning outcomes.

Dr Joel Suss

Joel has conducted research on performance in dynamic and complex environments, including wildfire fighting and law enforcement. He has employed a variety of cognitive task analysis techniques to examine expertise in these domains. Joel has recently completed his PhD using a prediction and option-generation paradigm to understand and improve decision making in law enforcement.

Professor Alexander J. Wearing

Alexander has taught and undertaken research in complex psychological processes. His recent work has involved collaboration with Lisa Frye as her PhD supervisor at the University of Melbourne, Australia.

Chapter 1
Human Factors in Emergency Management

Dr Christine Owen
Bushfire Cooperative Research Centre and University of Tasmania, Australia

Dr Peter Hayes
Bushfire Cooperative Research Centre and Kaplan Business School,
LaTrobe University, Australia

Introduction

People who manage emergency events face many challenges. This book focuses on the human factors challenges that are experienced by managers who deal with emergency events. Such events arise from hazards in the environment. Hazards can be natural, for example, earthquakes, wildfires, storms and tsunamis; they can be created by human activities, for example, oil spills and chemical explosions; or they can be intentional, for example, terrorism. Emergencies are actual or imminent events that pose a threat to life, property or the environment, and require a significant and coordinated response (EMA 1988). Some emergencies can be small in scale, like road accidents, while others can impact on thousands of people. A disaster has been defined as a crisis arising from significant damage, where agencies and the services they provide are overwhelmed by the damage and where the communities they work with are limited in their capacity to recover (Turner 1976). Although definitions of scale and complexity may be used to differentiate between emergencies and disasters, there are also some important qualitative differences. Quarantelli (2000) noted that compared with emergencies, disasters tend to require responding organisations:

- to work with more groups, many of which will be unfamiliar to those responding
- to have reduced autonomy
- to operate to different performance standards
- and to develop closer working arrangements between the public and private sectors.

Clearly these conditions present many challenges for those involved.

Emergency services are made up of people and technology-based systems for coping with adverse events, and the discipline of *human factors* can offer valuable insights into how these emergency services meet the challenges faced and how they can be continually improved. As a discipline human factors is concerned with 'understanding the interactions among humans and other elements of a system…[and] applies theory, principles, data and methods to design, in order to optimize human well-being and overall system performance' (Karwowski 2012, p. 3).

Most of the research discussed in this book has been undertaken with emergency services agencies responsible for either wildfire, and/or flood and storm management in Australia and New Zealand. Although personnel working within the fire and emergency management domain face a unique set of human factors challenges, the insights provided in this book will be of interest to researchers, scholars and practitioners involved in managing hazards of all kinds, such as technology and terrorism-based emergencies; emergencies in safety-critical industries like nuclear power, oil and energy production; and emergencies in health and transportation. This is because many of the challenges to be discussed in this book also can be found in these domains.

This book contributes to a growing body of knowledge about emergency services work and about the role of human factors involved in building capability in the sector. At the heart of this book are the following questions: how do personnel responding to emergency events manage these events, and what can be learned to enhance their capabilities?

The Growing Importance of Emergency Management

The emergency management sector performs a critical role in attempting to mitigate the harmful effects of a hazard that may develop into an emergency or even a disaster. Its activities therefore aim to protect individuals, communities, businesses, the economy and the environment from harm. In so doing, their purpose is to provide confidence to communities, governments and other stakeholder groups in the face of adversity.

In this respect the emergency services sector helps support the maintenance and development of stable and resilient communities. In many jurisdictions the part-funding of fire and rescue services by insurance companies is recognition of the loss minimisation role these emergency services agencies play in the economy. These agencies play a role in reducing the likely costs of emergency and disaster events, and, importantly, support resilient communities and businesses, so that they are able to recover more quickly from emergency events.

The emergency management sector comprises a variety of agencies, including fire and rescue, ambulance, police, coastguard, search and rescue, civil defence and state emergency services, as well as members of self-responding community groups. Emergencies often require several agencies to coordinate their response activities, e.g., fire, ambulance and police. Complex or large-scale events require multiple emergency services: various federal and state government departments;

local government and their municipalities; Red Cross and other non-government organisations; and logistical support from the military.

A core challenge for the emergency management sector is the fact that the number and intensity of adverse events is increasing and there is a growing vulnerability in our communities. This includes our social and ecological vulnerabilities, as illustrated by some of the major events that have happened globally in the past decade. These are outlined below:

- The 2004 Indian Ocean earthquake and tsunami had an impact on 14 countries where an estimated 275,000 people died. More than 600,000 people lost their livelihoods and 1.7 million were displaced.
- In 2010 a number of overwhelming disasters affected many populations. These included:
  - the earthquake in Haiti that affected an estimated 10 million people, creating one of the most complex urban disasters in decades (World Vision 2010); and
  - the record-breaking floods in Pakistan, which had an impact on almost all of the country. An estimated 20 million people were affected, and 1.89 million homes were damaged or destroyed (Red Cross 2010).
- In 2011 the Japanese earthquake, subsequent tsunami and nuclear power plant meltdown was reported as the toughest and most difficult crisis for Japan since the end of World War II (CNN 2011).
- The hurricanes and tornadoes now occurring across the Atlantic each year continue to cause considerable damage to countries such as Bermuda, Haiti, Jamaica, the Dominican Republic, the USA and sometimes Canada.

Australia and New Zealand have also had a series of significant emergency events and disasters that have tested and challenged governments and communities. The summer of 2010–2011 saw every State and Territory in Australia – except for the Australian Capital Territory – impacted by emergency events unprecedented in intensity and geographic spread. These occurred within two years of other exceptional events. The evidence gathered during the empirical research carried out for this book, and listed below, supports this claim:

1. **In 2009** Australia experienced its hottest month on record in January. In the state of Victoria, this heatwave was linked to 374 deaths (DHS 2009). In February of the same year, following a decade of drought, Victoria also experienced catastrophic bushfires where 173 people died and thousands of others remain displaced and multiple communities continue to recover.
2. **The 2010–2011 floods in Queensland and Victoria** were reported as Australia's wettest two-year period on record (BOM 2012). In terms of extent, impact and severity, the flooding along Australia's east coast was amongst the most significant in the country's recorded history.

3. **Tropical Cyclone Yasi** in February 2011 was one of the most powerful cyclones to affect Queensland and left behind significant damage, with a disaster declaration for a number of coastal and adjacent local government areas.
4. **The Christchurch earthquake** in February 2011 (itself one of 7,000 aftershocks to the September 2010 quake) caused major damage to half of the city centre's buildings, leaving many beyond repair (Brookings Institute 2012). In addition to the loss of 185 lives, this was the third largest insured event in history, triggering insurance claims of over $US12 billion.

Such large-scale, non-routine and overwhelming events have been called 'out-of-scale' events (Murphy and Dunn 2012). These writers argue that such events may now be considered as the 'new normal', meaning that they can no longer be considered as outlier aberrations or exceptions to the norm, but rather as indications of what can be expected on a more regular basis. The way we manage and recover from such events becomes increasingly important for a number of reasons, as summarised below.

Increasing Costs of Disaster

Perhaps not surprisingly, the costs of disasters are increasing. On average globally, there are approximately eight major natural disasters per year (IPCC 2012). From 1991 to 2005 around 60 per cent of costs due to disasters were incurred in OECD countries, most likely because of their higher asset base (Brookings Institute 2012). Although the number of disasters that occurred in 2011 was below average, it was the most expensive year in history in terms of disaster losses. These high costs were primarily because of the $210 (US) billion loss arising from the Japanese earthquake and the events that followed it (Brookings Institute 2012).

Increased Vulnerability

As world populations continue to grow, more people become vulnerable to the impacts of emergency events. In the 40-year period between 1970 and 2010, the world's population nearly doubled, from 3.7 billion to 6.9 billion, placing further pressures on arable lands, water catchments and infrastructure (UNISDR 2011). Moreover, drought and floods continue to put fragile ecosystems and developing countries at risk, and make the communities who live within them increasingly vulnerable. In the same 40-year period, for example, the average number of people exposed to flooding increased by 114 per cent – from 32.5 to 69.4 million annually, and mostly in non-OECD countries (UNISDR 2011). Compared to previous periods, an increasing number of people are now finding that they are living on flood plains with little or no economic alternative.

Changes in land-use practices, settlement patterns and industrial technology have also increased the vulnerability of communities to larger-scale emergencies.

This was demonstrated by the 2010 Deep-water Horizon drilling rig explosion in the Gulf of Mexico, and the 2011 Tohuku earthquake and tsunami that impacted the east coast of Japan and directly affected millions of people. In these examples, the Gulf of Mexico and the east coast of Honshū are heavily populated regions reliant on economic activities directly affected by the disasters, activities like fishing, aquaculture, agriculture and tourism. Furthermore, these events have very long tails, meaning that many communities may continue to be adversely affected for extended periods.

There is evidence that climate change may be also contributing to the intensity and duration of out-of-scale events. For example, climate change modelling suggests that the types of events outlined above are going to increase in their number, frequency and extent (IPCC 2012). In the case of wildfire, for example, drier and warmer conditions in Australasia, North America, Russia and Southern Europe mean that wildfire incidents are occurring with higher frequency and greater intensity. In addition, anticipated rising sea-levels, combined with increased storm and cyclone activity, represent an increasing hazard for those living on coastal areas or on low-lying coral atolls. People living in urbanised areas will also be affected, particularly by the growing interdependencies between, for example, energy, transport and agriculture (Boin and 't Hart 2010).

Increasingly Connected Systems

Our societies and their infrastructures have become increasingly inter-connected, e.g., water and transport systems. Other key systems, like food, have become more centralised. Because of the way these systems are organised, the impact of one out-of-scale event has wider implications, i.e. the impacts of a disaster experienced in one community can affect many others.

Increased interdependencies can also lead to other crises. A lengthy power outage in a major city during winter, for example, may not only put people at risk from hypothermia, but may create problems for managing traffic flows; it also may adversely affect business activity, and increase the risk of looting and other criminal behaviour.

A range of other factors such as demographic shifts, rural adjustment and broader business developments are also affecting community vulnerability and the delivery of emergency management services. Demographic changes, including changes to lifestyle expectations, domestic migration and community fragmentation, are increasing community vulnerability, as well as altering local social networks and the sustainability of volunteer groups (COAG 2011).

The increase in scale, intensity and duration of emergency events increases the need for cross-jurisdictional cooperation and support between the agencies responsible for managing hazardous events. Not only can an emergency event impact on several geographic areas, but the scale of these events may be such that additional resources are required from other states or even other countries. Recent international deployments of urban search and rescue (USAR) teams following earthquakes in New Zealand and Japan, and the exchange of wildfire personnel between Australia and North America to combat wildfires, highlights that emergency response is becoming an increasingly global issue in need of coordination.

In countries like Australia this sector relies heavily on volunteers and non-government organisations for prevention, preparedness, response and recovery activities. It employs the equivalent of over 97,000 full-time personnel and relies on more than 225,000 volunteers (SCRGSP 2012) as well as approximately 350,000 people involved in response and recovery activities (Howard 2009).

Moreover, the emergency services sector accounts for a significant component of central and local government expenditure in most countries. In Australia, for example, for the 2010–11 financial year, police, fire and ambulance services in Australia received over $11.3 billion of state and federal government funding.

Why the Emergency Services Environment is so Challenging

Working in the emergency management environment is extremely challenging for a number of reasons. Emergency events are dynamic and, at times, unpredictable as situations continue to evolve and change. Emergency services personnel may be required to operate under high levels of uncertainty and to make time-critical decisions using information that may be incomplete, inconsistent, or ambiguous, in part because the information available varies in quantity and quality. Emergency incidents are often a race against time, and personnel need to manage themselves well so that stress, fatigue and information-overload experienced by themselves and their colleagues do not undermine their performance. Responders are often required to work with community members who may be injured, traumatised or distressed by events. Emergency events can be particularly demanding for first responders and incident management teams.

Moreover, emergency events don't 'play by the rules' (Weick and Sutcliffe 2007). This contrasts with many other safety-critical domains where much of the work is procedural, and where safety is attained through collective mindfulness and adherence to well-established doctrine and protocols. These processes are important in the emergency services sector – however, the people responding to and managing emergency events have to also manage other layers of complexity.

Uncertainty, Unpredictability and Complexity

One of the key features that distinguish emergency services work from other safety-critical domains is that frequently, personnel are operating under 'degraded' conditions. Degraded conditions include: failures in critical equipment or technology; the required human resources are either unavailable or are overstretched or fatigued; and personnel are operating in hazardous (sometimes life threatening) conditions.

In other words, the hazards posed by emergency events mean that responders may be required to work in dangerous conditions that are unsafe, as well as unstable. For example, in order to get to injured or trapped people, responders may need to detour many kilometres around the path of a tornado, flood, or an area impacted by an earthquake. Ruptured gas mains, fallen power lines, and toppled trees and masonry can make streets both hazardous and difficult to navigate. Emergency events often have secondary impacts on the wider community that require careful management and often coordination with road, municipal, utility and humanitarian agencies – for example road closures, the requirement that community members stay indoors, or the provision of food and shelter to evacuees.

Different priorities may emerge and need to be managed as a result of such events. These priorities might be in conflict. Critical information that is needed might be missing. The time available to launch a response might be constrained and even compressed due to the interactions of other consequences of the events. Decisions that are made may lead to other unintended consequences. Many thousands of people and their livelihoods may be impacted.

Organising to Manage Emergency Events

In order to deal with the whole spectrum of emergency needs, considerable planning is required to bring together the normal endeavours of government and voluntary and private agencies in a comprehensive and coordinated way (EMA 1998; UK Cabinet Office 2013). International practice defines emergency management as the organisation and management of resources for dealing with prevention, preparedness, response and recovery .The United Nations Office for Disaster Risk Reduction (UNISDR) advises that the plans and institutional arrangements needed to engage and guide emergency management require the efforts of government, non-government, voluntary and private agencies to coordinate, in comprehensive ways, and respond to the entire spectrum of emergency needs (UNISDR 2011).

Successful emergency management, therefore, requires the cooperation and coordination of a diverse range of organisations. This includes the coordination of local, regional, state and national functions to support the communities affected. Typically, these activities have been divided into operational, tactical and strategic tasks where different types of demands and decisions are bounded, in part, by different time scales (ISO 22320, 2011; Paton and Owen 2013).

Defining roles and responsibilities
Table 1.1 below illustrates the different roles and responsibilities of first responders, the managers responsible for local incident tactics on the incident ground, and the roles and responsibilities of those involved in emergency management teams to strategically manage the longer-term consequences of response and recovery. As the table indicates, first responders are people who work directly on the fire

or incident ground. Personnel working at a local tactical management level operate in an incident management team. These teams develop plans and procure resources needed to manage the tactics on the incident ground. When an event is large in consequence or where there are multiple events, each with an incident management team, there is a need for overall coordination and management of the event's broader implications. This may occur at regional, state or national levels of government. Different emergency management roles require personnel to not only operate at different spatial scales, but also to operate at different temporal scales.

Table 1.1   Levels of coordination in emergency events

Layers of Emergency Management	Description	Australia/New Zealand Application
Operational	First responders; frontline personnel working directly on the fire or incident ground	First responders; incident ground personnel
Tactical	Local level incident management work directed at containing and mitigating the event	Incident management team
Strategic	Activity occurring above the local operational and tactical level that may involve regional, state or national activity. Concern for addressing the strategic issues across the whole of government and community	Regional/State National (New Zealand)

At a strategic level, personnel monitor local and tactical incident management operations in order to stay attuned to changes that may escalate events from the routine to larger out-of-scale ones. For strategic emergency management teams, the focus and the time frames will also be different because at this level there is a need for longer-term consideration of the direct and indirect consequences of the event, like the loss of tourism for a regional economy.

Incident management systems
Common incident management systems are central to managing the potential for large, complex, temporally challenging and dangerous incidents. In the United States, the National Incident Management System (NIMS) was developed following catastrophic wildfires in the 1980s where the need for better coordination was recognised (DHS 2008). Emergency management in Australia is governed by the Australasian Inter-Service Incident Management System (AIIMS) and in New Zealand it is called the Co-ordinated Incident Management System (CIMS). Both systems draw heavily on the structure of NIMS.

These types of incident management system are typically underpinned by three key principles, namely: management by objectives (all personnel involved in the incident work from a common set of objectives and complementary Incident Action Plans for achieving those objectives); functional management (which includes the utilisation of specific functions such as control, planning, operations, information and logistics) within the local incident management teams tasked with managing the incident); and the span of control (as an incident escalates, a supervising officer's span of control should not exceed five reporting groups) (AFAC 2013).

The incident management team is tasked with effectively and efficiently controlling the incident in terms of planning, resourcing, personnel management and safety, and its community/environmental impact (AFAC 2013). There is also a need to coordinate the functional responsibilities within the incident management team. These responsibilities are listed below.

1. Operations – responsible for operational resources and actions on the incident ground.
2. Planning – responsible for developing future plans and anticipating the resources that will be needed to meet those plans.
3. Intelligence – responsible for gathering information from disparate sources and making sense of this information.
4. Information – responsible for issuing community warnings and information to other stakeholder groups.
5. Logistics – responsible for procuring the various resources required to sustain the response effort (AFAC 2013).

An Incident Controller or Commander, and sometimes a Deputy, usually coordinates an incident management team. In addition to ensuring the incident management team functions effectively, this person has responsibility for informing or updating other agencies involved in the response like the police. This person also uses the media to brief other stakeholders and community members likely to be affected, as well as those involved in providing critical infrastructure, or those involved in local municipalities. To date, the systems based in Australian, New Zealand and the USA have been used mainly by agencies with fire and land management responsibilities, and by emergency services agencies responsible for natural disasters.

Human Factors in Emergency Management

Increasing amounts of data, greater specialisation, and the combination of increasing incident complexity and greater expectations of emergency managers all pose an enormous challenge for decision makers. Human factors are clearly central to effective performance in this team-based and technology-oriented environment. Personnel work in teams, as first responders, specialist units, or by undertaking command, control and communications (C3) activities. Emergency services (including local incident and other emergency management teams) are also becoming increasingly reliant on technology for collecting and disseminating information, operational communications, coordination and navigation.

Changes in information and communications technology mean that emergency agencies are faced with ever-increasing volumes and varying types of information (Klein 2009). Emergency management teams are now confronted with the challenge of quickly sorting and analysing this information, and using it in real time to inform their decision making for emergencies. Unfortunately, too much information can undermine decision-making performance (Omodei et al. 2005). Moreover, there appears to be increasing community expectations of emergency managers (O'Neill 2004). McLennan and Handmer (2011) observed, for example, that communities now expect up-to-date information on incidents such as wildfires and the timely provision of warnings. Recently several Australian emergency services agencies have developed and distributed incident information applications (or Apps) for smartphone users in the community.

Increased specialisation in the emergency services sector means that teams are increasingly made up of highly interdependent personnel, reliant on good communication for effective coordination and performance. Emergency services includes an increasing number of specialist teams. Some of these are made up of a wide range of expertise and have highly trained specialist personnel for particular roles. The Urban Search and Rescue task forces in Australia, for example, comprise engineers, paramedics, dogs and their handlers, rescue technicians, trauma doctors, hazardous materials specialists, logistics specialists and fire commanders (Mullins 2004).

Responding to larger incidents is a multi-agency response that often spans several jurisdictions like districts, counties, regions or states. A particular challenge with these larger events is the issue of inter-operability and coordination with agencies that may have to make trade-offs in decision making. For example, does a wildfire incident management team allocate its limited resources to protect valuable ecological natural habitat or to protect the local farmers' agricultural buildings? If resources are stretched in fighting a wildfire, should the emphasis be towards protecting a town, or a water catchment area that if damaged could take more than 100 years to recover and affect more people than those that live in the town?

Human Factors Challenges

The demands associated with incident complexity, increasing expectations, and the changes occurring within the industry, result in particular human factors challenges. According to a recent report by the Noetic Group (Murphy and Dunn 2012) the context now facing senior leaders in the field of public safety and emergency management is much more complex than that faced by their predecessors.

In some jurisdictions, personnel may be involved in their emergency services roles 24/7, in others, these responsibilities will be managed seasonally, alongside other substantive employment responsibilities. These extra responsibilities can add to the strain of trying to manage, in effect, two jobs. Many emergency services agencies rely on a mixture of paid and volunteer personnel. Often the volunteers provide important surge capacity during the initial response. However, if the incident timeline stretches to days and weeks, many of these volunteers will need to return to their day jobs, which then requires agencies to find replacement personnel.

Sometimes emergency management teams are formed on an ad hoc basis, where team members may or may not have previously worked together. Several chapters in this book address the issue of how team members work together. Although emergency services agency personnel generally receive common training, most teams tend to develop their own particular way of operating. Working alongside other personnel in challenging, high-stakes, time-critical conditions further adds to the complexity of the situation.

In addition to the human factors involved in emergency services operations, human factors are also relevant in understanding how to help communities prepare for emergencies, and how to communicate warnings and other prominent incident information for at-risk communities. The importance of carefully structured warning messages is crucial. Several chapters in this book touch on the issues of preparation and communication with affected communities.

Emergency services agencies undertake considerable preparation activities to ensure that the communities they serve and their own personnel are ready to respond to emergency events. Ongoing, high-level training exercises are required to prepare personnel to operate effectively in these complex environments. A difficult question for agencies is how many personnel need to be trained in each area and to what skill level? Emergency services agencies need to use a variety of training approaches, including simulation-based exercises, to develop and maintain the competencies of personnel. Developing simulation exercises of sufficient scale and complexity in order to train and prepare personnel is no easy task. Of course, to answer the questions of how many personnel should be trained and to what standard, requires agencies to first identify the possible threats that may pose a challenge and, thus, the level of capability and competency they need to build and maintain. Several chapters in this book provide insights that help address the challenges to do with training and resources.

Many of the above issues underscore the importance of the sector's leadership and capability, and the demands placed upon it. In addition to the resource and training challenges, there are the demands that emerge from greater community, government and political expectations. These newer demands require emergency managers to operate in a more open and political environment, to use more sophisticated strategic planning, and to put a greater emphasis on policy.

What the Reader Will Find

This book explores the challenges faced by those people involved in thinking, relating and coordinating emergency management work. The content is based on empirical research conducted at various levels of emergency management, including the incident ground, the local incident management teams, and the emergency management teams working at regional, state and national levels of coordination. It looks at the work done by personnel such as firefighters; incident management team members and team leaders, as well as coordinators of regional and state-level operations. How emergency services managers and vocational learning instructors support enhanced emergency management response capability also receives attention.

The first three chapters are focused on thinking processes and in particular the strategies first responders use when making decisions on the fire ground. The process of thinking in emergency response is constrained by the time-critical, stressful and dynamic conditions involved in mitigating or managing emergency events. This is particularly the case for people who are first-on-scene or working on the incident ground. Events, such as bushfires, are continually unfolding, and this can place communities and firefighters at risk. Sometimes biases can influence thinking which can then influence decision making. This book examines the potential sources of bias and identifies strategies that can be used to support the thinking processes of those responders, so that they get the best outcomes possible under challenging conditions.

The way in which stress influences thinking processes is discussed by Jim McLennan, Roger Strickland, Mary Omodei and Joel Suss and in Chapter 2. It provides an overview of the literature on stress and its impact on decision making. They make the point that high levels of stress may compromise firefighter, safety-related decisions and actions on the fire ground in at least four ways: (a) attention is likely to become narrowly focused on only a few aspects of the developing situation, so that emerging threats may not be acted upon; (b) important tasks may take longer than anticipated and mistakes may be more likely; (c) working memory is likely to be impaired and important information may not be remembered; and (d) forming sound judgements and making good decisions may become progressively more difficult as thinking becomes more rigid. In their chapter they discuss the implications of these insights for capability development and suggest particular approaches to training for survival modes, as well as for a need to increase awareness about the impact of stress on thinking. They also suggest that the impact of stress on thinking processes needs to be taken into account in any subsequent inquiry, and any learning from near misses and mistakes.

Claire Johnson, in Chapter 3, focuses on the expert decision-making processes of experienced firefighters and their use of thinking about worst-case scenarios when planning their actions. Worst-case scenario thinking involves identifying possible worst-case outcomes, so that strategies can be enacted to reduce the probability of such a negative outcome. In high-risk work environments, these kinds of strategies are particularly important for supporting reliability and avoiding accidents and even death. However, as Johnson observes, such thinking can limit decision making if not well managed. Her chapter discusses the biases that can sometimes influence decision making and outlines a process that has emerged, through her research, for improving the effectiveness of worst-case scenario thinking. Such processes can be used by individuals, as well as by team leaders, to support effective decision making and shared mental models between team members.

The level of self-awareness or metacognition about thinking processes is discussed by Lisa Frye and Alex Wearing in Chapter 4. They discuss the role that self-awareness plays when working under the time-critical, dangerous and pressured situations of firefighting. They used *think aloud* protocols during simulations with a group of experienced and volunteer firefighters to document the errors made by personnel when overloaded. Their work also identifies some of the metacognitive strategies that experienced personnel use in order to better manage under these conditions. These insights enabled Frye and Wearing to identify strategies that can be used to assist firefighters to regulate their thinking in different types of situation and in particular to develop strategies so that firefighters may be better able to manage decision trade-offs, that is, when one action needs to be traded off against another because of competing goals. Their chapter provides some useful resources that will be of interest to practitioners aiming to enhance metacognitive (or self-regulation) skills for personnel involved in safety-critical work.

Chapter 5 is the first of four chapters examining teamwork in emergency management response. Almost all emergency management work involves people relating to one another in teams. This makes the issues of communication and teamwork critically important. Stress and the high consequences involved in responding to emergencies can pose challenges to effective communication and teamwork. Sometimes conflicting goals need to be managed and negotiated.

Aspects that influence how emergency management team members relate to one another, like the way they manage their emotions, is discussed by Jan Douglas in Chapter 5. It provides a different perspective on the role of emotion in emergency management teams. Acknowledging the important role of stress, Douglas also examines the ways in which team members experience a range of emotions and how these permeate both thinking and interpersonal relating within and between teams. In her qualitative study she interviews incident management team members who describe feelings of joy and confidence as well as despair and fear. These emotions are also interpreted collectively and become the basis for positive and negative stereotypes, which in turn influence collective sense-making within teams. The way in which emotion influences positive or negative team performance is then discussed. The implications for practitioners and facilitators of learning include promoting cultural awareness, as well as emotional self- awareness.

In Chapter 6 Peter Hayes takes up the issue of familiarity between team members and the potential impact this has for team coordination, communication and trust. The reported findings suggest that where familiarity existed, teams performed better than the teams in which there was unfamiliarity or mixed familiarity between members. Their performance was affected in terms of the workload carried by the team and the quality of the communication. In the teams with familiar members, two intra-team communication measures were higher: listening to other team members and providing constructive comment to other team members. This indicates that psychological safety was an important aspect of team climate and that this positively influenced team performance. Peter makes some suggestions for emergency services managers so that they may be able to bring out the best in their teams and some suggestions for training interventions to improve performance.

Leadership, communication and teamwork is discussed by Christine Owen in Chapter 7. The importance of psychological safety is also taken up in this chapter. Both Owen and Hayes draw on research by Lewis, Hall and Black (2011) who observe that even experienced firefighters may face social pressures to remain silent in some situations because they do not feel psychologically safe to express concern about danger. Owen observed 18 incident management teams at work and the influence of the team leader on the communications climate of the team. She compares observations of the communication styles of two groups of team leaders: those perceived as effective and others perceived as less effective by their team members as well as subject matter experts. The more effective team leaders engaged in higher levels of coaching activity. This involved establishing a more open communication styles that facilitated more effective intra-team and inter-team communication. These she calls boundary riding (where the team leader coaches team members to be mindful of temporal constraints), boundary spanning (where the emphasis is on ensuring within team function integration and synchronisation of activities) and boundary crossing (where the emphasis is on managing external expectations and relationships). In undertaking these three forms of coaching activity team leaders facilitated more effective within- and between-team coordination.

External relationship management is important because responding to major emergency events requires the coordination of a range of different groups and teams operating at different levels in an emergency management arrangement. These include those involved in mitigating and/or managing the threat, like response teams and organisations; those involved in recovery, like community services or humanitarian agencies and teams; and those involved with the restoration of services like transportation, food, electricity and communications supply. The way information flows between various teams is of particular importance in emergency management. The challenges of coordination are an important focus of this book and are discussed in Chapter 8. In this chapter Peter Bremner, Chris Bearman and Andrew Lawson trace the challenges of coordinating decisions made at local, regional and state levels of emergency management. As noted previously in this introduction, firefighters and incident commanders make tactical decisions about emergency fire management at the local incident level. At regional and state levels fire officers make more broadly strategic decisions to coordinate resources with other stakeholders. These officers working at regional and state levels of emergency management coordination also have a responsibility to ensure that the levels below them are performing effectively, and the authors outline the kinds of decisions that need to be made at these different levels in emergency management. They also discuss the implications of social pressures that can impede decision making. These pressures can lead to errors and, ultimately, breakdowns in coordinated decision making.

The final two chapters consider how insights gained from both research and experience may be coordinated into new learning to enhance capability.

Coordinating insights, so that teams and organisations can adapt to future possible scenarios, is also a significant challenge in the emergency services domain. The role of training to enhance thinking and to improve team communication and team performance is given attention by Ben Brooks in Chapter 9. He takes up the challenges of investigating what can be done within emergency services organisations to address the possibilities of errors in decision making, and on a positive note, how training may support emergency services personnel to develop capability. This requires greater attention to developing skills in complex problem solving, and in leadership and communication. He also observes the effect of degraded conditions in emergency management response, i.e. many of the usual supports that help people to do their work are not functioning effectively. Emergency services workers then need to make do, even if they are fatigued or technologies are not working properly, or the needed resources are not available. Under these circumstances a new framework for operation is needed and Brooks calls this entering the 'Zone of Coping Ugly'. He concludes by identifying strategies that can be used in training to support coping under extreme conditions.

Finally, how individuals, teams and organisations can learn by reflecting on emergency events is discussed by Sue Stack in Chapter 10. She explores the challenges for emergency services organisations in creating cultures of reflective inquiry and learning, where people have the opportunity to explore ideas, re-frame their understandings and speak honestly about their emotions and concerns. She draws on an organisational learning concept in use in the US wildfire community called the 'Staff Ride'. The Staff Ride offers a structured opportunity where reflective inquiry is possible by setting out guidelines for conversation within a situated learning environment. In Sue Stack's chapter she looks at two Staff Rides recently organised in Australia. She considers how they foster reflection, the quality of reflection, the impact on participants and how engagement in the design and reflection of a Staff Ride provides deeper insights about potentials for reflective organisational learning. The chapter concludes with some suggestions for how practitioners may employ this organisational learning approach to facilitate critically reflective learning cultures and how they may learn from experience in emergency events.

Conclusion

The chapters in this book fill an important gap in knowledge about human factors in the fire and emergency services industry. Emergency services coordination and response is growing in importance – both in Australia and world-wide. Yet there is very little by way of published human factors books addressing the audience of this industry.

The research reported in this book is original and identifies important areas of performance that are critical to the safety and reliability of personnel involved in fire and emergency services. The discussions about thinking, relating and coordinating are also pertinent to others working in similar high-reliability, high-consequence domains. The chapters therefore contribute to an integrated body of work about individual and group performance and their limitations.

Emergency services response work is not as easily bounded as in other safety-critical domains. This is because of the distinct features of emergency events which involve wicked problems for which there may be unanticipated consequences and highly interdependent consequential effects. These research chapters, therefore, tell us something about a potential future world of work that is highly dynamic, interdependent and for which improvisation and critical thinking and problem solving are necessary prerequisites.

Finally, given the indications of increases in extreme weather-related events, the need for a better understanding of human factors in the fire and emergency services domain has never been more timely or needed. It is for these reasons, that this book represents an important contribution to this emerging field.

Acknowledgements

The research was supported by a Bushfire Cooperative Research Centre Extension Grant. However, the views expressed are those of the authors and do not necessarily reflect the views of the Board of the funding agency.

References

Australasian Fire and Emergency Service Authorities Council (AFAC) (2013). *The Australasian Inter-service Incident Management System (AIIMS)*. Fourth edition, 2013 revision.

Boin, A. and 't Hart, P. (2010). Organising for effective emergency management: Lessons from research. *Australian Journal of Public Administration*, 69, 357–371.

BOM (2012). National Climate Centre, Bureau of Meteorology; Australia's wettest two-year period on record; 2010–2012. *Special Climate Statement.*

Brookings Institute, The (2012). The year that shook the rich: a review of natural disasters in 2011 *London school of Economics project on internal displacement* London: The Brookings Institute

CNN (2011). Japanese PM: 'Toughest' crisis since World War II. http://www. worldvision.com.au/Issues/Emergencies/PastEmergencies/HaitiEarthquake/ ACallToAction_Haiti6Months.aspx [Accessed Archived from original on 12 April 2011. Retrieved 12 March 2012].

COAG (2011). *National strategy for disaster resilience: Building the resilience of our nation to disasters.* ACT: Commonwealth of Australian Governments.

DHS (Department of Homeland Security) (2008). *National incident management systems.*

DHS (Department of Human Services) (2009). *Heatwave in Victoria: An assessment of health impacts, January 2009.* Melbourne, Government of Victoria.

EMA (1998). Australian Emergency Management Glossary. *Australian Emergency Manuals, Series Part I: The Fundamentals Manual 3.* Commonwealth Government of Australia.

Howard, B. (2009). Climate change and the volunteer emergency management sector. *National Emergency Response,* Winter 2009, 8–11.

IPCC (2012). Managing the risks of extreme events and disasters to advance climate change adaptation. In Field, C.B., V. Barros, T.F. Stocker, D. Qin, D.J. Dokken, K.L. EBI, M.D. Mastrandrea, K.J. Mach, G.-K. Plattner, S.K. Allen, M. Tignor and P.M. Midgley (eds) *A Special Report of Working Groups I and II of the Intergovernmental Panel on Climate Change.* Cambridge, UK, New York, USA: Cambridge University Press.

Karwowski, W. (2012). The discipline of human factors and ergonomics. In Salvendy, G. (ed.) *Handbook of human factors and ergonomics* (4th edition). Hoboken: Wiley (pp. 3–37).

Klein, G. (2009). *Streetlights and shadows: Searching for the keys to adaptive decision making.* Cambridge: MIT Press.

Lewis, A., Hall, T.E. and Black, A. (2011). Career stages in wildland firefighting: Implications for voice in risky situations, *International Journal of Wildland Fire,* Vol. 20, pp. 115–124.

McLennan, B. and Handmer, J. (2011). Framing challenges for sharing responsibility: A report of the sharing responsibility project. Melbourne: RMIT University and Bushfire Cooperative Research Centre

Mullins, G. (2004). Urban search and rescue: developing Australia's capability. *The Australian Journal of Emergency Management,* 19, 6–9.

Murphy, J.G. and Dunn, W.F. (2012). Medical education in the clouds. *CHEST Journal,* 141.

O'Neill, P. (2004). *Developing a risk communication model to encourage community safety from natural hazards.* Sydney: NSW State Emergency Service.

Omodei, M., McLennan, J., et al. (2005). More is better?: A bias towards overuse of resources in naturalistic decision-making settings. In Montgomery, H.,

R. Lipshitz and B. Brehmer (eds) *How professionals make decisions*. Mahwah: Erlbaum (pp. 29–41).

Patton, D. and Owen, C. (2013). Incident management. In Penuel, P., M. Statler and R. Hagen (eds) *Encyclopedia of crisis management*. Thousand Oaks, CA: Sage.

Pickrell, J. (2005). Facts and figures: Asian Tsunami Disaster. *New Scientist.* Retrieved on 30 July 2013 from: http://www.newscientist.com/article/dn9931-facts-and-figures-asian-tsunami-disaster.html.

Quarantelli, E.L. (2000). *Emergencies, disasters and catastrophes are different phenomena*. Preliminary paper no. 304, Disaster Research Center, University of Delaware.

Red Cross (2010). Pakistan floods: Deluge of disaster. Retrieved on 31 July 2013 from: http://reliefweb.int/report/pakistan/pakistan-floodsthe-deluge-disaster-facts-figures-15-september-2010.

Rogers, L. (2010). Why did so many die in Haiti's quake? *BBC News*. Retrieved on 30 July 2013 from: http://news.bbc.co.uk/2/hi/americas/8510900.stm.

SCRGSP (2012). *Report on Government Services 2012* (Vol. 1). Melbourne, Steering Committee for the Review of Government Service Provision (SCRGSP) and Productivity Commission.

Turner, B.A. (1976). The development of disasters: a sequence model for the analysis of the origins of disasters. *Sociological Review*, 24, 753–774.

UK Cabinet Office (2013). Improving the UK's ability to absorb, respond to, and recover from emergencies. Retrieved on 31 August 2013 from: https://www.gov.uk/government/policies/improving-the-uks-ability-to-absorb-respond-to-and-recover-from-emergencies.

UNISDR (2011). Global assessment report on disaster risk reduction. The United Nations Office for Disaster Risk Reduction. Retrieved on 31st August, 2013 from: http://www.unisdr.org/we/inform/publications/19846.

Weick, K.E. and Sutcliffe, K.M. (2007). *Managing the unexpected: Resilient performance in an age of uncertainty*. San Francisco, CA: John Wiley & Sons, Inc.

Whittaker, J. and Handmer, J. (2010). Community bushfire safety: a review of post-Black Saturday research. *Australian Journal of Emergency Management*, 25, 7–13.

World Vision Australia (2010). A call to action: Haiti at 6 months. Retrieved 31 July 2013 from: http://www.worldvision.com.au/Issues/Emergencies/PastEmergencies/HaitiEarthquake/ACallToAction_Haiti6Months.aspx.

Chapter 2
Stress and Wildland Firefighter Safety-related Decisions and Actions

Prof. Jim McLennan

School of Psychological Science, La Trobe University, Victoria, Australia
and Bushfire Cooperative Research Centre, Victoria, Australia

Roger Strickland

Country Fire Authority (CFA), Victoria, Australia

Dr Mary Omodei

School of Psychological Science, La Trobe University, Victoria, Australia
and Bushfire Cooperative Research Centre, Victoria, Australia

Dr Joel Suss

Department of Cognitive and Learning Sciences, Michigan Technological
University, Michigan, USA

Introduction

Both official reports (e.g., United States Department of Agriculture Forest Service 2001) and informal accounts (such as those by Maclean 2003) indicate that stress is often associated with wildland firefighter safety-compromising incidents. However, the processes by which stress may compromise safety in threatening situations on the fireground are seldom discussed. In this chapter we examine findings from the stress and human performance research literature concerning fear/anxiety and cognitive[1] abilities, and discuss their relevance for understanding the behaviour of firefighters under imminent threat from wildfire.

Here, we use the term *stress* to refer to the totality of an individual's negative psychological experiences associated with a wildfire threat as the stressor: fear and anxiety in particular; but also worry, frustration and anger. Other potential stressors discussed in this book include: fatigue and heat (see Bremner, Bearman and Lawson, Chapter 8, this volume); operational responsibility (see Douglas, Chapter 5, this volume); and complex team interactions (see Brooks, Chapter 9, this volume). We follow Lazarus and Folkman (1984) in conceptualising such

---

1 We follow generally accepted practice within psychology by using the term *cognitive* to refer to human perception, learning, remembering, thinking and acting (for example Sternberg 2009, p. 578).

psychological stress as resulting from interactions among three elements of an individual's appraisal of a wildfire threat situation under conditions of uncertainty: (i) the likely actions demanded of the individual; (ii) the individual's self-perceived ability to cope with these demands; and (iii) the perceived severity of threat posed by the stressor (Figure 2.1). We distinguish *stress*, as a negative mood state, from *arousal* as activation or energisation (Cox and Mackay 1985). For a discussion of alternative conceptual accounts of stress in relation to human performance see Staal (2004).

We acknowledge that individuals differ, both in tendencies to experience negative mood states such as anxiety and fear in the face of threats, and in the kinds of impacts their fear and anxiety might have on safety-related decision making; however, space constraints do not allow us to explore this (see Staal et al. 2008).

Method

The research literature concerned with stress and human performance is extensive. Numerous studies have examined effects of different stressors on cognition and performance, including fatigue, heat and cold, hunger and thirst, noise, sleep deprivation, time pressure, mental workload, and physical and psychological threats. These studies were conducted in a range of settings, including laboratory studies, simulation exercises, naturalistic environments such as combat training, and field settings such as parachuting and underwater diving. We first examined general reviews, including those by Driskell and Salas (1996), Hammond (2000),

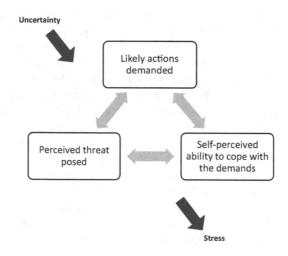

Figure 2.1   The Transactional Model of Stress (Lazarus and Folkman 1984)
Source: Reprinted with permission

Staal (2004), Kavanagh (2005) and Staal et al. (2008). We then made electronic searches of data bases (such as PSYCHINFO, Web of Science, Google Scholar) for experimental or quasi-experimental quantitative studies of performance in the face of potentially life-threatening stressors, in field or naturalistic settings.

The criteria for inclusion in our review were:

1. the research used an experimental or quasi-experimental design
2. the stressors were likely to be perceived as having the potential to cause pain, injury or death
3. cognitive performance was assessed in a field or other naturalistic setting
4. the cognitive performance assessment tasks had to be related directly to the stressful aspects of the task environment rather than being incidental to the task environment.

The methodological importance of the final criterion was discussed by Berkun (1964). Only 10 reports of experimental research conducted in field or naturalistic settings (such as parachuting, diving, rock climbing and military training) were located which examined effects of stressors generating fear or anxiety associated with threat of possible pain, injury or death, on performance of cognitive tasks.[2] In the absence of systematic research involving wildfire safety-related decision making specifically, we used these plus an additional laboratory study[3] to infer likely effects of imminent wildfire threat on firefighter cognitive performance.

Review of the Literature on Stress and Performance

General reviews of the stress and performance literature indicate that, overall, stress of all kinds tends to degrade cognitive processes and impair performance. Table 2.1 summarises the studies we identified as potentially relevant to firefighters making safety-related decisions and taking actions under imminent wildfire threat: the 11 reports described a total of 17 separate studies. Most reported evidence that the task situation was stressful, as measured by increase in heart rate, self-reported increase in anxiety following exposure to the stressor, or biological assay (Ice and James 2006). More details of the studies are in McLennan et al. (2011). We followed the suggestions of Staal (2004) to examine findings in relation to four aspects of cognition: attentional control; perceptual-motor skills; memory; and judgment and decision making.

---

2   We excluded studies, like that by Lieberman et al. (2005), which employed stressors such as sleep deprivation, fatigue and hunger in addition to physical safety threat.

3   We included a study by Keinan (1987). While it was conducted in a laboratory, it fulfilled the other criteria, used a self-report measure to confirm that participants in the stress conditions reported higher levels of stress than controls, and the findings have been cited previously as evidence for the negative effects of stress on decision making.

Table 2.1   Fear/anxiety-related performance decrements: participants; stressor; potential implications for wildfire safety

% Mean Decrement (Study)	Participants	Stressor	Potential Implications for Firefighters under Imminent Wildfire Threat
Attentional Control			
28% reduction in maintaining mental focus on the primary task (Kivimaki and Lusa 1994)	Young adult male firefighters	Dark, potentially hazardous environment to be navigated wearing breathing apparatus	Difficulty in maintaining concentration on an essential task
48% narrowing of attentional focus (Pijpers et al. 2006)	Novice rock climbers	Height	Attention is focused narrowly on the main potential threat, so other emerging hazards may not be attended to
31% narrowing of attentional focus (Weltman and Egstrom 1966)	Novice divers	Open ocean dive	Attention is focused narrowly on the main potential threat, so other emerging hazards may not be attended to
50% narrowing of attentional focus (Weltman et al. 1971)	Students in a pressure chamber	Simulated air pressure equivalent to 60 feet (18.3m)	Attention is focused narrowly on the main potential threat, so other emerging hazards may not be attended to
Perceptual Motor Skills			
6% reduction in manual dexterity (Baddeley and Idzikowski 1985)	Novice divers	Imminent open-water dive	Difficulty in assembling, handling or operating equipment, clumsiness
23% slower visual letter-search time; 40% slower symbol-search time (Idzikowski and Baddeley 1987)	Novice parachutists	Imminent parachute jump	Slow to notice changes in the environment
5% slower choice-reaction times (Jones and Hardy 1988)	Female university students	Imminent 4.6m jump down to padded mats	Slow to respond to sudden changes in a threat situation

Table 2.1   *Continued*

72% slower climbing time, associated with more, tentative, limb movements (Pijpers et al. 2006)	Novice rock climbers	Height	Complex action sequences slower to complete successfully
Memory			
38% reduction in retrieval of survival information from long-term memory (Berkun, 1964)	Army recruits in basic training	Anticipated emergency aircraft 'ditching'	Difficulty in remembering important safety-related information
19% reduction in working memory capacity (Idzikowski and Baddeley 1987)	Novice parachutists	Imminent parachute jump	Difficulty in remembering a correct sequence of actions needed to successfully undertake complex tasks
41% reduction in working memory capacity; 11% reduction in processing efficiency (Leach and Griffith 2008)	Novice and experienced parachutists	Imminent parachute jump	Reduced ability to retain new information, to interpret the significance of changes in the environment and to remember a correct sequence of actions needed to complete complex tasks
33% reduction in working memory efficiency (Robinson et al. 2008)	Nautical college students	Helicopter underwater emergency evacuation training	Reduced ability to interpret the significance of changes in the environment
Reasoning, Judgement and Decision Making			
10% reduction in reasoning and judgement accuracy (Berkun 1964)	Army recruits in basic training	Anticipated emergency aircraft 'ditching'	Errors in interpreting information; making safety-compromising decisions
33% reduction in reasoning and judgement accuracy (Berkun 1964)	Army recruits in basic training	Fear of being accidentally shelled by artillery	Errors in interpreting information; making safety-compromising decisions
18% reduction in reasoning and judgement accuracy (Berkun 1964)	Army recruits in basic training	Fear of having accidentally injured a fellow-soldier in a demolition exercise	Errors in interpreting information; making safety-compromising decisions

Table 2.1   *Concluded*

7% reduction in reasoning performance (Idzikowski and Baddeley 1987)	Novice parachutists	Imminent parachute jump	Difficulty in interpreting warnings and processing information in the environment so as to form an accurate assessment of the threat and matching this to safety options
39% reduction in reasoning performance, associated with incomplete and inefficient scanning of decision alternatives (Keinan 1987)	University students	Threat of electric shock if incorrect alternative selected	Hasty, incomplete and inefficient consideration of safety options

Attentional Control

Findings from four studies indicate that maintaining concentration on a primary task becomes more difficult (Kivimaki and Lusa 1994), and attentional control is degraded by anxiety so that perception becomes more narrowly fixated, or tunnelled (Weltman and Egstrom 1966; Weltman, Smith and Egstrom 1971; Pijpers et al. 2006). The implications are that firefighters experiencing high levels of anxiety due to imminent wildfire threat may: (a) find it difficult to concentrate on tasks central to safety, and (b) fail to attend to cues of newly emerging dangers – such as radio information, wind changes or approaching vehicles.

Perceptual Motor Skills

In six studies which employed a variety of perceptual-motor tasks there was a decrement in performance associated with increased anxiety: changes in the task environment took longer to detect (Jones and Hardy 1988; Idzikowski and Badderley 1987) and movements were slower and less skilful (Badderley and Idzkowski 1985; Pijperset al. 2006). Implications are that firefighters experiencing high levels of anxiety due to the stress of imminent wildfire threat may: (a) be slow to detect some threatening developments in their environment; (b) take somewhat longer than normal to complete complex tasks; and (c) be less skilful, or clumsier, than usual in some of their actions – such as deploying personal fire shelters.

Memory

Findings from four studies suggest that high levels of anxiety are likely to reduce both working memory capacity (Idzikowski and Badderley 1987) and processing efficiency (Leach and Griffith 2008; Robinson et al. 2008), and to interfere with retrieval of knowledge from long-term memory (Berkun 1964). The implications are that individuals under imminent wildfire threat may have difficulty in: (a) keeping safety-relevant procedures in mind; (b) correctly interpreting the significance of emerging threats; and (c) being able to remember safety-enhancing information – such as escape routes.

Reasoning, Judgement and Decision Making

Findings from five studies indicated that high-level cognitive processes involved in reasoning, judgment and decision making are degraded by stress: there were decrements in the ability to apply logic in reasoning (Berkun 1964; Idzikowski and Badderley 1987) and in the ability to fully and efficiently consider decision options (Keinan 1987). The implications are that individuals experiencing high levels of anxiety due to imminent wildfire threat may engage in safety-compromising behaviour by: (a) misunderstanding information provided to them; or (b) failing to evaluate relevant options available before committing to an action.

Stress and Performance

The unweighted mean cognitive performance decrements associated with stress in the studies reviewed (Table 2.1) were:

1. perceptual motor skills: 29 per cent
2. attentional control: 39 per cent
3. memory functioning: 33 per cent
4. reasoning, judgement and decision making: 21 per cent
5. across all 17 studies: 30 per cent.

These findings from the stress and performance research literature suggest the potential for significant negative effects of fear/anxiety on cognitive performance. Firefighters experiencing high levels of fear/anxiety due to imminent wildfire threat as a stressor may make decisions and take actions which jeopardise their safety. On the fireground some stressed firefighters might: (a) become distracted from essential tasks and fail to notice cues of emerging threats; (b) be slow to respond to indications of threat and be clumsy in their actions; (c) have difficulty keeping safety issues in mind and remembering important information; and (d) find it hard to think issues through so as to select and implement a safe option.

Evidence from the Fireground

Two objections can reasonably be raised against a too-ready application of the above findings from the research setting to the fireground. First, with the possible exceptions of the research by Berkun (1964) and by Kivimaki and Lusa (1994), the cognitive tasks were artificial and unrelated to the kinds of tasks that firefighters undertake on the fireground. Second, with the possible exception of the study by Kivimaki and Lusa, the participants in the research had no prior training or experience on the tasks – unlike wildland firefighters. If the discussion of research summarised Table 2.1 is to have merit beyond an academic exercise then there should be evidence of firefighter stress-related cognitive performance decrements from the fireground.

In an attempt to respond to the objections we looked for evidence of stress-related cognitive performance decrements associated with safety-compromising wildland fire incidents. We asked colleagues to describe safety-compromising incidents where stress may have played a role. We then examined reports describing wildland firefighter fatalities, injuries and near misses from two sources: the Wildland Fire Lessons Learned Centre web site (http://www.wildfirelessons.net/ Home.aspx) and Centres for Disease Control National Institute of Occupational Safety and Health Firefighter Fatality Reports (http://www.cdc.gov/niosh/fire/). We were able to locate several fireground incident investigation reports where it seemed very likely that firefighter stress played a role; however, we noted that few contained direct evidence that those involved were, in fact, experiencing high levels of stress at the time. It seemed likely that in several investigations the possible contribution of stress to the adverse events in question had not been considered explicitly.

A sample of potentially relevant incidents is summarised in Table 2.2. These should be regarded as suggestive of ways in which stress *may* degrade cognitive performance and compromise safety on the fireground. The incidents illustrate how stress-related impairments of attentional control processes, perceptual-motor skills, memory – both working memory and knowledge retrieval – and judgement and decision making can play a role in compromising wildland firefighter safety.

Table 2.2   Possible firefighter stress-related cognitive failings associated with safety-compromising decisions and actions

Cognitive Function	Example (Source)
Attentional control	When multiple spot fires broke out and smoke impeded visibility two lookouts became apprehensive. They decided to leave the lookout site and move to the safety zone. They began a rapid retreat down a rocky gully. Lookout #1 did not take time to put on his gloves, believing their lives were in danger. Subsequently, while attempting to break in to a building to shelter from the oncoming fire he sustained injuries to his hand requiring medical treatment (Lessons Learned Review, Horseshoe 2 Fire) A firefighter died as a result of being struck or run over by a fire vehicle. The victim was trying to escape from a fire that was about to overrun his fire vehicle. The victim left the vehicle and attempted to escape on foot. Ten other fire vehicles were attempting to leave the area. Due to the volume of vehicles attempting to escape the fire, the poor visibility due to the smoke, as well as a variety of tanker mechanical issues resulting from the intense heat, firefighters were driving their tankers off the road, frequently colliding with each other, blocking each other in, and, as a result, many had to escape the fire by backing out (NIOSH Firefighter Fatality Report F2011-09)
Perceptual-motor skills	As the tanker began to be burned-over, the driver (so he believed) deployed the vehicle protection water spray system. The tanker was destroyed, the crew narrowly escaped death. The spray system operating switch was later found to be in the 'off' position (CFA Upper Ferntree Gully tanker entrapment, based on the crew's account)
Memory	When the wind changed there was a blow-up. In their haste to escape an imminent burnover, the crew forgot to depressurise the hose and they could not disconnect it. This meant that as they tried to reverse the fire truck to safety there was a risk that the hose would foul the wheels. Fortunately the hose burnt through. (CFA Upper Ferntree Gully tanker entrapment, based on the crew's account)
Judgement and decision making	Two firefighters (DIVS1 & DIVS2) were on foot when their escape was cut off by the fire. DIVS1 said, 'We need to deploy! We need to deploy!' They discarded their packs and started to deploy their fire shelters. DIVS1 & 2 raked the ground to prepare the site for deployment. DIVS2 looked at the surrounding fuels and decided that a shelter deployment would not be survivable. He said 'We need to go down the hill!...We won't survive with deployment.' Finally he said, 'Follow me! I am going down the hill!' DIVS1 did not respond. DIVS2 balled up his shelter under his arm and ran down the hill. DIVS1 had deployed his shelter and did not follow. Subsequently, DIVS2 reached a road safely. DIVS1 did not survive. (USDA FS Accident Investigation Report Panther Fire Entrapment)

Countering the Adverse Effects of Stress on Cognitive Performance

Incidents involving high levels of imminent threat during a wildfire need not necessarily cause stress-related cognitive performance deficits sufficient to result in serious injury or death. The following is an account of wildland fire crew survival during a tanker entrapment and burnover:

> The fire was now crowning in the trees on the windward side of the tanker. The crew rolled down the cab inner reflective safety curtains and velcroed the curtain edges together. Conditions in the cab were now extremely hot and pitch black. The front passenger window glass broke, probably due to direct flame impact. The velcro strips fastening the safety curtain flaps melted, forcing the crew leader to hold them together with his gloved hands. Unable to drive forward for fear of hitting the tanker in front, the driver attempted to reverse back along the track. The view to the rear was almost completely obscured by smoke and the roll-down reflective safety curtain. The tanker veered off the track, collided with a large tree and was immobilised. The crew attempted to retrieve woollen blankets from cab lockers to cover themselves but had great difficulty opening the sealed plastic stowage bags wearing their wet gloves. Embers entered the cab through the broken window and ignited the plastic dashboard. The crew leader extinguished this fire temporarily with his feet. One of the crew began to express fear, and was silenced authoritatively by the crew leader. As parts of the tanker began to burn and conditions in the cab became marginal, the crew leader who was sitting on the windward side of the truck, asked the rear seat crew member to check the state of fire on the lee side. This was judged to be now survivable. The crew exited the cab of the burning tanker successfully after one of the crew assisted the crew leader out over the cab centre console by hauling on his coat, then cushioning the crew leader's fall to the ground with his body. All three then walked out of the burning forest together, shielding themselves from radiant heat using a single woollen blanket. Injuries were limited to minor burns and bruising. (Country Fire Authority, Upper Ferntree Gully tanker entrapment, 23 February 2009; based on the crew's account)

In-depth interviews with the crew by the second author suggested that the key to their survival was their ability to: (a) control their level of stress sufficiently to be able to maintain attentional focus on survival-relevant aspects of the threat environment; (b) identify and make sound judgements about survival options; and (c) take effective and timely survival-enhancing actions. This analysis is consistent with other accounts of survival under extreme threat (e.g., Wise 2009; McLennan et al. 2011), which emphasise the importance of down-regulating stress levels to permit sound judgements and effective actions. There has been considerable research about psychological processes involved in emotion regulation generally (for a review see Koole 2009) and a critical factor in down-regulating fear/anxiety appears to be prior training or experience relevant to survival in the particular threat situation (Wise 2009).

While we are not aware of research which shows specifically that training enhances firefighter safety under imminent wildfire threat, there is abundant research evidence from the training literature, especially military training, of the benefits of training for subsequent effective actions under stress (e.g., Keinan, Friedland and Sarig-Naor 1990; Freidland and Keinan 1992; Zach, Raviv and Inbar 2007; Lukely and Tepe 2008; Aguinis and Kreiger 2009; McClernon et al. 2011). Taber (2010) cited research reporting overall survival rates from helicopter crashes in the ocean of 92 per cent for those with helicopter underwater escape training, and only 66 per cent for those without such training.

Concluding Discussion

In his classic 1832 treatise *On War*, Clausewitz asserted 'Everything is very simple in war, but the simplest thing is difficult' (1982, p. 164). Our review suggests that in wildland firefighting, threat-induced stress can make the simplest decisions and actions difficult, because stress can degrade any or all of four cognitive functions: attention, perceptual-motor skills, memory, and judgement and decision making. This has important implications for several aspects of wildland fire safety. We suggest three in particular: training, operations, and after-action reviews and investigations.

For wildland firefighter safety training, survival mode procedures need to be simple, and be practised frequently under realistically simulated life-threat conditions. The purpose of such training is to generate protective behaviours which are relatively automatic and do not require high levels of mental workload and complex thinking, which can be disrupted easily by high levels of anxiety due to threat. Training unit personnel must keep in mind a fundamental distinction (Anderson 1983) between recognitional knowledge (knowing about) and procedural knowledge (knowing how to). They must resist cost pressures to substitute classroom recognitional knowledge acquisition about safety in place of hands-on learning and practising safety exercises and drills which are as realistic as possible. In about AD75, the historian Flavius Josephus wrote of the Roman approach to waging war: '…it would not be far from the truth to call their drills bloodless battles, and their battles bloody drills' (1959, p. 195). Perhaps agencies should approach survival training for their wildland firefighters in the same spirit!

For operations, *worst case scenario* thinking (see Johnson 2011; and Chapter 3, this volume) needs to be cultivated in order to foster anticipation of rapid escalations of threat situations. 'Plans are best-case scenarios. Let's avoid anchoring on plans when we forecast actual outcomes. Thinking about how the plan could go wrong is one way to do it' (Kahneman 2012, p. 128). Safety-related scenarios should be a central component of training activities, using all available methods, including table top map exercises, staff rides and simulations. The importance of self-monitoring generally, and especially under threat, should be emphasised so that firefighters become skilled at recognising indications that they are experiencing increased stress levels and may need to take effective emotional self-regulation actions (see also Brooks, Chapter 9, this volume). Such actions can include: positive self-talk, mental re-evaluation of the situation and assumptions, information- or advice-seeking, physical movement, calming breathing, and effortful-ignoring of issues which are not immediately relevant to the most serious emerging threats (Seaward 2009; Australian Psychological Society 2012; US Marine Corps 2012).

For after-action debriefs and reviews, and post-incident investigations, possible stress-related decrements in judgement and decision-making quality need to be taken into account. This means first that potential stressors, stress experiences, and effects on judgements, decisions and actions should be acknowledged, discussed and explored. For investigators, the possibility of stress-related degradation of cognitive performance should be a key human factors issue to be considered and analysed carefully rather than, say, being viewed simply as evidence of human failure.

We noted in the course of our literature searches that researchers have given relatively little systematic attention to effects of stress on wildland firefighter performance. It seems highly desirable that researchers increase our knowledge and understanding of the effects of threat-related stress on wildfire safety through studies focused on firefighters' experiences and actions on the fireground. As two pioneers of disaster research observed:

> In the laboratory one can produce very frightening (and even traumatic) experiences but must stop short of those which constitute a real threat to the continued existence and health of the subjects involved and actual disasters do not stop short. An experiment cannot introduce the disaster stresses of overwhelming threat to life and limb…(Fritz and Marks 1954, p. 26)

Acknowledgements

The research was supported by a Bushfire Cooperative Research Centre Extension Grant. However, the views expressed are those of the authors and do not necessarily reflect the views of the Board of the funding agency. We thank Claire Johnson (Country Fire Authority Victoria, see Chapter 3, this volume) and Damien Killalea (Tasmania Fire Service) for helpful comments on a previous draft. The chapter is based on a paper presented to the 12th International Wildland Fire Safety Summit in Sydney, October 2012.

References

Aguinis, H. and Kreiger, K. (2009). Benefits of training and development for individuals and teams, organizations and society. *Annual Review of Psychology*, 60, pp. 451–474.

Anderson, J.R. (1983). *The architecture of cognition*. Cambridge, MA: Harvard University Press.

Australian Psychological Society (2012). *Psychological preparedness for natural disasters*. Melbourne: Australian Psychological Society. Available at www. psychology.org.au/publications/tip_sheets/disasters/

Baddeley, A.D. and Idzikowski C. (1985). Anxiety, manual dexterity and diver performance. *Ergonomics*, 28, pp. 1475–1482.

Berkun, M.M. (1964). Performance decrement under psychological stress. *Human Factors*, 6, pp. 21–30.

Berkun, M.M., Bialek, H.M. et al. (1962). Experimental studies of psychological stress in man. *Psychological Monographs: General and Applied*, 76(15), Whole Number 534 1962, pp. 1–39.

Clausewitz, C. (1832, 1982). *On war*. London, Penguin.

Cox, T. and Mackay, C. (1985). The measurement of self-reported stress and arousal. *British Journal of Psychology*, 76, pp. 183–186.

Driskell, J.E. and Salas, E. (eds) (1996). *Stress and human performance*. Mahwah, NJ: Lawrence Erlbaum.

Friedland, N. and Keinan, G. (1992). Training effective performance in stressful situations: Three approaches and implications for combat training. *Military Psychology*, 4, pp. 157–174.

Fritz, C.E. and Marks, E.S. (1954). The NORC studies of human behavior in disaster. *Journal of Social Issues*, 10(3), pp. 26–41.

Hammond, K. R. (2000). *Judgments under stress*. New York: Oxford University Press.

Ice, G.H. and James, G.D. (2006). *Measuring stress in humans: A practical guide for the field*. Cambridge: Cambridge University Press.

Johnson, C. (2011). How bushfire fighters think about worst case scenarios. *Fire Note* #77 March 2011. Melbourne, Bushfire Cooperative Research Centre. Available at: http://www.bushfirecrc.com/managed/resource/worst_case_scenarios.pdf.

Idzikowski, C. and Baddeley, A.D. (1987). Fear and performance in novice parachutists. *Ergonomics*, 30, pp. 1463–1474.

Jones, J.G. and Hardy, L. (1988). The effects of anxiety on psychomotor performance. *Journal of Sports Sciences*, 6, pp. 59–67.

Josephus, F. (*circa* AD75, 1959). *The Jewish war*. London: Penguin.

Kahneman, D. (2012). *Thinking, fast and slow*. London: Penguin.

Kavanagh, J. (2005). *Stress and performance: A review of the literature and its applicability to the military*. RAND Corporation, Technical Report 192. Santa Monica. Available at http://www.rand.org/pubs/technical_reports/TR192.html.

Keinan, G. (1987). Decision making under stress: Scanning of alternatives under controllable and uncontrollable threats. *Journal of Personality and Social Psychology*, 52, pp. 639–644.

Keinan, G., Friedland, N. and Sarig-Naor, V. (1990). Training for task performance under stress: The effectiveness of phased training methods. *Journal of Applied Social Psychology*, 20, pp. 1514–1529.

Kivimaki, M. and Lusa, S. (1994). Stress and cognitive performance of fire fighters during smoke diving. *Stress Medicine*, 10, pp. 63–68.

Koole, S.L. (2009). The psychology of emotion regulation: An integrative review. *Cognition and Emotion*, 23, pp. 4–41.

Lazarus, R.S. and Folkman, S. (1984). *Stress, appraisal, and coping*. New York: Springer.

Leach, J. and Griffith, R. (2008). Restriction in working memory capacity during parachuting: A possible cause of 'no pull' fatalities. *Applied Cognitive Psychology*, 22, pp. 147–157.

Lieberman, H.R., Bathalon, G.P. et al. (2005). The fog of war: Decrements in cognitive performance and mood associated with combat-like stress. *Aviation, Space, and Environmental Medicine.* 76(7-Supplement), pp. C7-C14. Available at: http://www.ncbi.nlm.nih.gov/pubmed/16018323.

Lukely, B.J. and Tepe, V. (eds) (2008). *Biobehavioural resilience to stress*. Boca Raton: CRC Press.

Maclean, J.N. (2003). *Fire and ashes: On the front lines of American wildfire*. New York: Henry Holt & Company.

McLennan, J., Omodei, M. et al. (2011). Bushfire survival-related decision making: What the stress and human performance literature tells us. In R.P. Thornton (ed.) *Proceedings of the Bushfire CRC & AFAC 2011 Conference Science Day*, pp. 307–319, Melbourne, Bushfire Cooperative Research Centre. Available at http://www.bushfirecrc.com/managed/resource/307-319_bushfire_survival-related_decision_making.pdf.

McLennan, J., Omodei, M. et al. (2011). 'Deep survival': Experiences of some who lived when they might have died in the February 2009 bushfires. *Australian Journal of Emergency Management*, 26(2), pp. 41–46.

McClernon, C.K., McCauley, M.E. et al. (2011). Stress training improves performance during stressful flight. *Human Factors*, 53, pp. 207–218.

Pijpers, J.R., Oudejans, R.R.D. et al. (2006). The role of anxiety in perceiving and realising affordances. *Ecological Psychology*, 18, pp. 131–166.

Robinson, S.J., Sunram-Lea, S.I. et al. (2008). The effects of exposure to an acute naturalistic stressor on working memory, state anxiety and salivary cortisol concentrations. *Stress*. 11, pp. 115–124.

Seaward, B.L. (2009). *Managing stress: Principles and strategies for health and well-being* (6th edition). Boston: Jones & Bartlett Publishers.

Staal, M. (2004). *Stress, cognition, and human performance: A literature review and conceptual framework*. Ames Research Centre, NASA/ TM–2004–212824.

Moffett Field. Available at http://human-factors.arc.nasa.gov/flightcognition/Publications/IH_054_Staal.pdf.

Staal, M., Bolton, A.E. et al. (2008). Cognitive performance and resilience to stress. In B.J. Lukely and V. Tepe (eds), *Biobehavioural resilience to stress*, pp. 259–300. Boca Raton: CRC Press.

Sternberg, R.J. (2009). *Cognitive psychology.* Belmont: Wadsworth.

Taber, M. (2010). *Offshore helicopter safety report.* Prepared for the Offshore Helicopter Safety Inquiry Newfoundland and Labrador-Canada. Canada-Newfoundland and Labrador Offshore Petroleum Board, Volume 2 – Expert and Survey Reports, pp. 211–290. Available at http://www.cnlopb.nl.ca/pdfs/ohsi/ohsir_vol2.pdf.

United States Department of Agriculture, Forest Service (2001). *Thirtymile Fire investigation: Accident investigation factual report and management evaluation report.* Washington DC. Available at http://www.fs.fed.us/t-d/lessons/documents/Thirtymile_Reports/Thirtymile-Final-Report-2.pdf.

US Marine Corps (2012). *Combat and operational stress control.* MCRP 6-11C Washington DC, US Navy. Available at http://www.marines.mil/Portals/59/Publications/MCRP%206-11C%20%20Combat%20and%20Operational%20Stress%20Control.pdf.

Weltman, G. and Egstrom, C.H. (1966). Perceptual narrowing in novice divers. *Human Factors*, 8, pp. 499–505.

Weltman, G., Smith, J.E. and Egstrom, C.H. (1971). Perceptual narrowing during simulated pressure chamber exposure. *Human Factors*, 13, pp. 99–107.

Wise, J. (2009). *Extreme fear: The science of your mind in danger.* New York: Palgrave Macmillan.

Zach, S., Raviv, S. and Inbar, R. (2007). The benefits of a graduated training program for security officers on physical performance in stressful situations. *International Journal of Stress Management*, 14, pp. 350–369.

Chapter 3

Expert Decision Making and the Use of Worst Case Scenario Thinking

Dr Claire Johnson

Country Fire Authority, Victoria, Australia

Introduction

The consideration of possible worst case scenarios (WCS) is critical in the planning and anticipation that is required for effective decision making when managing an emergency event. This chapter describes a programme of research that explored the characteristics of expert decision making in an emergency management setting and identified the key barriers that can interfere with long-term planning and consideration of WCSs.

Defining Worst Case Scenarios

The research reported in this chapter explored how experienced firefighters engage in WCS thinking when managing and mitigating an emergency event. Consistent with the definitions of Sunstein (2007) and Clarke (2006), WCSs are defined as hypothetical future events that (1) have extreme negative outcomes and (2) are novel, rare or unlikely (i.e., non-routine). WCS thinking, or planning, involves identifying possible worst case outcomes so that action can be taken to reduce the probability of that scenario occurring or the severity of the outcome if it does occur. WCS planning originated in the military domain, particularly during the Cold War, where it referred to planning for possible nuclear war outcomes (Fahey and Randall 1998).

A good example of WCS thinking is defensive driving: if you coast through an intersection with your foot resting on the brake in case a car incorrectly turns in front of you, then you have prepared for a possible WCS. However, one driver might anticipate a car turning incorrectly, whereas another might consider and prepare for the possibility of their brakes failing. As the perception of likelihood and consequence is subjective, the events that individuals identify as WCSs will vary.

Previous research identified WCS thinking as important in the processes of prediction and planning that are essential for effective decision making (McLennan, Omodei, et al. 2007). The benefits of WCS thinking include:

1. reducing the chance of being surprised by unexpected events

2. highlighting faulty assumptions and errors in decision making
3. identifying actions to reduce the probability of a WCS event occurring
4. identifying actions to mitigate the severity of consequence if WCS events cannot be avoided.

WCS thinking, however, is a critical skill that is easily overlooked and can be difficult to execute. Nevertheless, failure to consider these potential events can lead to tragic consequences.

Implications of WCSs in Emergency Management

Failures to effectively manage potential WCSs have been specifically implicated in a number of investigations of high-profile fatality fire incidents in Australia. For example, during an inquiry into bushfires in the Australian Capital Territory in January 2003, in which four lives were lost, authorities were reported as displaying an attitude of 'dogged optimism...[and the] tendency to view the situation from a "best-case scenario" perspective' (McLeod 2003, p. 62). An independent report into the Wangary bushfire in South Australia during January 2005, in which nine lives were lost, recommended that those managing a large fire 'should be continually reminded by prompts in the system to plan and resource for "worst case scenario"' (Smith 2005, p. 70). The findings of these two investigations emphasised the importance of WCSs in real-world decision making and the potential for tragic outcomes if WCSs are not sufficiently managed.

Such findings have parallels internationally. For example, after describing numerous fatal fire events in the United States, Desmond (2007) noted that 'high-risk organizations might avoid accidents if they resist underestimating risk and instead overestimate threats; that is, they should attempt to imagine "worst case scenarios", regardless of their rarity, and plan accordingly' (p. 341, note 313). In another book concerned with tragic firefighter deaths, Alexander and colleagues (2007) directed fire practitioners to 'plan for and expect the worst every time out' (p. 45). The continuation of loss of life of both firefighters and community members reinforces the need for continual improvement in anticipating WCSs (Keller 2004). While extreme fires occur infrequently, the consequences are so significant that all fire organisations must do more to improve the management of WCSs.

The Cognition of WCS Thinking

WCSs can severely challenge the capacity of bushfire fighters to make effective decisions because of the innate limits and tendencies of the human brain. In complex, dynamic and uncertain situations, such as firefighting, research has shown that experts rely on experienced-based intuitions to make high-quality decisions (e.g., Beach and Lipshitz 1993; Klein 1998; Gigerenzer 2007). Intuitive decisions are based on the skilled application of automatic processes developed through extensive experience. Intuitions allow decision makers to automatically use their experience base to rapidly understand situations and take action (Klein 1998; Kahneman and Klein 2009). When based on extensive and appropriate experience in a domain with predictable cues, intuitive decisions can be fast, accurate, practical and effective.

However, such experience-based intuitions can be vulnerable to error, particularly when required to manage novel situations. All decision makers have a limited set of experience, which reduces the effectiveness of intuitive decisions when situations develop in unfamiliar ways (Klein 1998). The advantages of expertise-based decision making are only valid if the current situation does not differ too far from previous experience, and expert skills only transfer to tasks that are similar to tasks that have previously been performed (Proctor and Vu 2006). The low frequency of WCSs makes them particularly difficult to predict, imagine and plan for because decision makers have not usually experienced such situations previously. These limits of experience-based intuitions have critical implications in the bushfire domain because the majority of decision makers would not have had direct experience with the most destructive bushfire events.

Not only are WCSs challenging because decision makers cannot rely on experience to produce fast and effective intuitive decisions, but innate cognitive tendencies often make WCSs appear less likely (see also Frye and Wearing, Chapter 4, this volume). Unconscious cognitive shortcuts, or heuristics, provide decision makers with quick and simple rules to guide judgements and decisions (Gigerenzer and Todd 1999). While heuristics are often effective in many situations, they sometimes lead to errors in judgements and predictions. For example, the availability heuristic refers to the tendency to make predictions of event frequency based on how easily an example of that event can be brought to mind. Over (2007, p. 6) described an application of the availability heuristic:

> Suppose we are trying to make a judgment about whether to trust a man who is smiling at us. We might recall many people who smiled at us and were trustworthy, and only a few who smiled as us and were not. We are using the availability heuristic (Tversky and Kahneman 1973) for probability judgment if we conclude, from these memories, that it is highly probable that the man is trustworthy.

If the availability of memories is approximately similar to the actual probability of smiling men being trustworthy, then this heuristic will result in a satisfactory decision (Over 2007). If not, the available memories may result in a bias, possibly leading to an incorrect decision or error.

Numerous factors influence cognitive availability, including familiarity, salience and emotional connection (Tversky and Kahneman 1973, 1974; Finucane et al. 2000; Slovic et al. 2004; Boehm and Brun 2007). Events are typically more cognitively available if they are more recent (i.e., recency effect; Frensch 1994) or easier to imagine (i.e., simulation heuristic; Kahneman and Tversky 1982). While personal experiences are particularly influential, availability of vicarious experiences can approximate likelihood when no personal experiences exist (e.g., training examples, media reports and peer storytelling). Availability could lead to a biased prediction of the likelihood of WCS events if recall of past experiences (personal or vicarious) does not approximate actual likelihood.

In addition to an availability bias, unrealistic optimism may influence predictions of the likelihood of WCS events. Optimism bias describes the tendency to underestimate the likelihood of negative events and overestimate positive events (Weinstein 1980). For example, laboratory studies conducted by Armor and Taylor (2002) found that 'between 85 per cent and 90 per cent of respondents claim that their future will be better...than an average peer' (p. 336). Similarly, the overconfidence effect describes inaccurate confidence in a judgement; people tend to overrate their abilities in a range of activities and skills (e.g., Adams and Adams 1961; Fischhoff, Slovic and Lichtenstein 1977; Russo and Schoemaker 1992; Hoffrage 2004). People also tend to underestimate difficulties of action implementation; the planning fallacy describes unrealistic optimism concerning the risks, time and cost of implementing a plan (Buehler, Griffin and Ross 1994). Overall, these biases of unrealistic optimism produce a tendency to underestimate the likelihood of dangerous or unpleasant outcomes, including WCS events, and overestimate the ability to manage such situations.

Therefore, cognitive biases can challenge effective decision making to manage WCS events because they exacerbate the inevitable limits of experience and prompt decision makers to underestimate the likelihood of non-routine situations with potential for severe outcomes. Underestimating the probability of WCS events could severely interfere with decision making because the future situation and relevant risks will not be accurately understood. This risk underestimation can also substantially distort or disrupt processes of prediction, forecasting and planning. If biases cause WCSs to be considered unlikely, then decision makers are more likely to be surprised when WCS events do occur and less likely to have the capacity to respond to unfolding disasters. Risk underestimation has considerable implications in the fire domain, where it can lead to faulty decision making and has been identified as a factor in many firefighter deaths (Alexander, Mutch and Davies 2007). Thus, the majority of decision makers in bushfire situations are likely to find the management of WCSs challenging because of the limits of human cognition.

To counteract these negative effects on risk perception, adaptive experts develop metacognitive skills that allow them to more flexibly manage unexpected events and WCSs. In a highly informative book, *Building expertise: Cognitive methods for training and performance improvement*, Clark (2008) defined metacognition as 'the skill that sets goals, plans an approach to accomplish a goal, monitors progress towards the goal, and makes adjustments as needed along the way' (p. 314). As described in more detail in Frye and Wearing (Chapter 4, this volume), metacognitive skills have a range of benefits, including improving decision outcomes in non-routine situations where experienced-based decisions are not available (Cohen, Freeman and Thompson 1995). Metacognition allows decision makers to assess the accuracy of situational awareness, evaluate potential solutions, monitor the influence of biases, review decision effectiveness, and manage stress and emotional arousal (Klein 1997, 1998; Pliske, McCloskey and Klein 2001; McLennan, Pavlou and Omodei 2005; Feltovich, Prietula and Ericsson 2006).

Research has found that requiring decision makers to perform metacognitive processes of critiquing decisions and considering alternative decision outcomes can reduce some biases (e.g., Larrick 2007; Milkman, Chugh and Bazerman 2009). For example, Arkes (1991) found decision makers had reduced levels of overconfidence after they were asked to 'consider the opposite' of the option they were about to choose. Koriat, Lichtenstein and Fischoff (1980) found participants who described contradicting reasons for their choices displayed less overconfidence when answering dual-alternative history or geography questions (e.g., Kiev is the capital of (a) Ukraine or (b) Romania). Augmenting decision making with metacognitive skills of self-reflection and critique can help to avoid the limits of experience, reduce bias, minimise errors, and improve problem-solving. Along with a wide variety of experience, strong metacognitive skills are a key requirement for the development of adaptive expertise.

The skills of adaptive experts, including metacognition, have been identified as important in contributing to high safety records in High Reliability Organisations (HROs). High Reliability Organisations are organisations that encounter fewer disasters than would reasonably be expected in domains characterised by extreme, uncertain and dynamic conditions. In *Managing the unexpected: Resilient performance in an age of uncertainty*, organisational psychologists Weick and Sutcliffe (2007) described the characteristics of HROs and expounded their widely accepted view that the ability to manage unexpected events with high consequences (i.e., WCSs) differentiates these organisations from others with less impressive safety records.

Weick and Sutcliffe (2007) investigated the management practices of various organisations they identified as high reliability, which included nuclear power plants, aircraft carriers, hospital emergency departments, airforce squadrons and air traffic control towers. The authors identified five common characteristics that enable greater safety in hazardous settings:

1. Preoccupation with failure: Decision makers deliberately concentrated on possible failures, which made failures more cognitively available and reduced unrealistic optimism.
2. Reluctance to simplify: Decision makers were better able to establish accurate expectations because they understood that (a) the working environment was complex, (b) assumptions and labels were potentially damaging, and (c) simplifications could conceal latent problems.
3. Sensitivity to operations: Decision makers paid attention to frontline processes, which allowed them to notice anomalies and weak signals of impending unexpected events.

4. Commitment to resilience: Decision makers managed errors more effectively and rapidly because they acknowledged the inevitability of mistakes, envisioned undesired consequences, and practised the skills required in extreme events.
5. Deference to expertise: Decision makers desired the most accurate knowledge from the most well-informed source and, rather than adhering to hierarchy, allowed decisions to migrate to those with the greatest knowledge.

The importance of effectively managing WCSs seemed to have a direct link to the impressive safety reports of HROs, even though they exist within challenging environments (Weick and Sutcliffe 2007). Under extreme time-pressure and cognitive load, the personnel of HROs relied on experience-based intuitive processes. However, where possible, intuitive decision making was augmented with more deliberate analytic processes (e.g., critiquing) because of the limits of experience and the potential impact of critical human errors (e.g., availability bias, optimism bias). Along with using multiple decision strategies, HRO personnel developed adaptive expertise by being sensitive to feedback, developing metacognitive skills, and gaining a wider variety of past experiences. They imagined how the situation could deteriorate or how plans could fail and mentally simulated the possible actions to manage those outcomes (Weick 1995; Weick and Sutcliffe 2007). In situations of high time-pressure and dynamic conditions, decision makers coped with unexpected outcomes (i.e., WCSs) as effectively as cognitively possible.

The characteristics of HROs also had disadvantages because they (a) required considerable experience and training to develop, (b) had time, cost and cognitive capacity tradeoffs, and (c) were challenging to implement in real-world domains. However, the value of this robust HRO research was that it provided a gold standard for how organisations should operate to ensure high reliability and effective management of WCS events. Weick and Sutcliffe (2007) described a range of behaviours displayed by personnel in organisations with highly successful safety records, which reflected the research purpose which was to guide the management of more conventional organisations by identifying organisational practices of HROs. Nevertheless, Weick and Sutcliffe did not progress deeply into identifying the underlying cognitive processes of successfully managing WCS events; further effort is required to expand understanding in this area.

Previous Research Examining WCSs

While extensive psychological research has been conducted into real-world decision making, the topic of WCSs has received little direct research attention. Previous research highlighted the importance of considering WCSs (e.g., McLennan et al. 2007; Weick and Sutcliffe 2007) and provided some insight into the role of WCSs in real-world decision making (e.g., Orasanu and Fischer 1997; McLennan, Pavlou and Omodei 2005), including my previous research of WCSs in bushfire decision making (Johnson 2005). However, previous research failed to provide a conceptual understanding of the role of WCSs or clarify how WCS events might influence real-world decision making.

Given the potential consequences of extreme events, such lack of explicit research attention on the topic of WCSs in the psychological literature presented a considerable gap that required further research attention. Research into WCSs had the potential to improve psychological understanding of how WCSs are involved in decision making and to extend current theoretical conceptions of human real-world decision making. Such research was likely to have implications for real-world decision making in a wide range of similar domains.

The research described in this chapter explored the role of WCSs in bushfire decision making in a research programme of three sequential interview studies (Johnson 2011; Johnson, Cumming and Omodei 2007a, 2007b, 2008). The research programme aimed to:

1. investigate how Australian bushfire fighters think about WCS events
2. discover what factors might prevent decision makers from identifying or preparing for such events
3. identify methods with the capability to improve decision outcomes in relation to WCS thinking.

Method

Study 1 examined WCS thinking by paid and volunteer bushfire fighters at all levels of experience (i.e., non-expert sample) and all levels of command (i.e., from crew leader in charge of one fire truck to incident controller responsible for an entire fire). I analysed 54 interviews that had previously been conducted at routine (non-problematic) bushfires. Interviews were carried out as soon as possible (usually within 24 hours) using the Human Factors Interview Protocol (HFIP; Omodei, McLennan and Reynolds 2005). No interview questions specifically asked about WCSs.

Study 2 examined the WCS thinking of domain experts (i.e., highly experienced bushfire decision makers). Thirty highly experienced personnel (volunteer and paid staff) from five Australian fire agencies were identified by peers and recruited for interviewing. Interviews in Study 2 differed from Study 1 because they:

1. targeted domain experts with extensive experience in managing major bushfire incidents
2. explicitly asked interviewees about WCSs
3. focused on the incident management level of command, where decision makers were involved in managing large complex incidents

4. discussed non-routine incidents of a critical nature that interviewees felt had challenged their decision making skills and required all their expertise to manage
5. considered incidents that had occurred during the previous five years.

The Worst Case Scenario Interview Protocol (WCSIP) was specifically developed for the purposes of this study, based on the Critical Decision Method (CDM) (Crandall, Klein and Hoffman 2006).

Study 3 focused on gaining feedback from fire agencies about the validity of my research findings and exploring potential methods for improving WCS thinking. Semi-structured interviews were conducted with eight highly experienced training and operational improvement personnel. Interviewees reported that the results of the previous two studies successfully reflected the decision making carried out and the challenges faced during bushfire incidents.

Worst Case Scenario Thinking

The analysis of interview data from all three studies followed a style orientated towards discovery and open-mindedness suggested by Crandall, Klein and Hoffman (2006). Stages of analysis were fluid because multiple passes through the data maximised insights and enabled systematic examination of emerging issues.

In the analysis of interviews from Study 1, I coded discrete sections of text (excerpts) into preliminary categories as they were identified. Excerpts ranged from a sentence to a number of paragraphs, depending on how long the interviewer discussed the issue. Excerpts were coded into multiple categories if they related to more than one issue. Some categories had been identified in the literature review as important concepts, while others became evident only as coding continued. In this way, the development of coding categories was a combination of top-down and bottom-up processes, which allowed themes to emerge from the data. A coding scheme was developed to define and guide coding of primary themes. This coding scheme was then applied to all interview data. This section integrates the results from the three empirical studies exploring WCS thinking.

WCS Reporting

Analysis indicated that most interviewees (83 per cent in Study 1 and 90 per cent in Study 2) reported considering WCSs, to some extent, when recounting a past incident. However, the reporting of WCS thinking was highly variable because some interviewees did not report any WCSs and others reported several (up to 10 per interview).

The majority of interviewees reported that WCS thinking was a critical aspect of incident management. Interviewees reported that bushfire fighters must consider:

1. multiple scenarios (e.g., worst case and most likely)
2. multiple timeframes (e.g., next few hours and next few days)
3. multiple perspectives (e.g., a detailed micro view and a 'big picture' macro view).

WCS Content

All WCSs reported by interviewees in both Study 1 and 2 focused on one of three topics:

1. a possible outcome causing harm to the community (e.g., civilian fatality)
2. a possible outcome causing harm to firefighters (e.g., firefighter fatality)
3. a possible outcome causing the fire to become more difficult to contain and control (e.g., escaping a containment line).

WCS content varied between decision makers at different roles in the command structure. For instance, low-level decision makers (for example, crew leader) were more likely to report WCSs with a firefighter focus, whereas high-level decision makers (for example, incident controller) were more likely to report WCSs with a community or fire focus. However, interviewees emphasised that the ultimate WCS would involve the fatality of someone under their command.

Possible Barriers and Expert Strategies

Analysis identified a wide range of factors with the potential to interfere with accurate consideration of WCSs. The factors identified as potential barriers to WCS thinking were:

1. underestimation of risk, time or space (e.g., underestimation of the risk of a burnover, or the distance a fire will travel)
2. tunnel vision (e.g., fixation on one area of the fire)
3. lack of appropriate experience (e.g., no grass fire experience)
4. lack of local knowledge (e.g., limited knowledge of local terrain)
5. suboptimal attitudes to risk and safety (e.g., assumption of a best-case outcome)
6. situational characteristics (e.g., uncertainty of information)
7. interpersonal issues (e.g., tensions between locals and non-locals)
8. standard procedures (e.g., inflexibility of planning procedures).

Interview responses suggested that even expert incident managers could sometimes overlook these important elements of planning and be affected by barriers to WCS thinking. However, the domain experts interviewed in Study 2 were aware of these potential barriers and had developed strategies to minimise their influence on effective decision making.

A range of behaviours and strategies enabled experts to effectively identify and prepare for WCSs. Many expert strategies reported in Study 2 were consistent with behaviours identified within those five High Reliability Organisations characteristics. Table 3.1 summarises the strategies for reducing barriers to WCS thinking reported by domain experts and indicates the corresponding HRO characteristics (numbers in parenthesis indicate relevant HRO characteristics).

Table 3.1   List of expert strategies to minimise barriers to WCS thinking, with indications of the corresponding HRO characteristic

Potential Barrier	Expert Strategies
Underestimation of risk, time and distance	Check plans and assumptions with others (HRO 2) Assume the worst and plan accordingly (HRO 1) Develop a timeline or consider forward spread on a map Identify critical times or tasks necessary for future planning (HRO3)
Tunnel vision	Take a time-out from the incident management team (IMT) for a mental break Delegate tasks and responsibilities Establish a clear command structure Set up additional structures or processes to enable long-term planning (HRO 2, 3 and 5)
Lack of appropriate experience	Use scenario exercises in training (HRO 4) Mentor developing decision makers (HRO 4) Target training to particular individuals Debrief after successful and unsuccessful incidents (HRO 4) Take and make opportunities to learn (HRO 4) Investigate near-miss events (HRO 1)
Lack of local knowledge	Retain locals in the IMT to provide assistance (HRO 5) Integrate local and non-local crews (HRO 5)
Suboptimal attitudes to risk and safety	Establish a team culture of questioning (HRO 2) Ask others to critique plans and check assumptions (HRO 2) Establish a formal group process of discussing and critiquing plans (HRO 2)
Situation characteristics: escalation, lack of coordination and control, lack or uncertainty of information, and lack of firefighting resources	Gather information as early as possible (HRO 3) Find accurate sources of information (HRO 3) Implement unambiguous command structures Establish clear responsibilities and reporting procedures Request additional firefighting resources as early as possible Prioritise the areas or tasks that most require firefighting resources

Table 3.1   *Concluded*

Potential Barrier	Expert Strategies
Situation characteristics: time-pressure, stress and fatigue	Use relaxation techniques Slow decision processes to maximise accuracy Double-check plans and calculations with others (HRO 2)
Interpersonal issues	Establish pre-formed management teams (HRO 5) Train in groups or teams to get to know the skills of others (HRO 5) Be aware of, and proactively manage, internal and external pressures Identify fundamental safety priorities so pressures do not undermine planning
Standard procedures	Audit effectiveness of IMT structure based on current and future need (HRO 3 and 4) Adjust procedures and structures where necessary (HRO 3 and 4)

Note: Brackets represent the corresponding High Reliability Organisation (HRO) characteristic from Weick and Sutcliffe (2007).

In summary, most of the expert strategies identified could be categorised into the following themes:

1. *'what if' thinking* – experts questioned the situation and imagined how things could go wrong
2. *back-up plans* – experts developed a number of back-up plans to deal with a range of eventualities, with trigger points indicating when to change plans
3. *self-management* – experts reflected on their own thought processes and took action to manage cognitive overload or stress
4. *critiquing plans* – experts encouraged respectful discussion and dissent to highlight faulty assumptions and check decision processes
5. *adaptive decision making* – experts audited the effectiveness of management structures and adapted standard procedures if required
6. *focus on fundamentals* – experts focused on fundamental rules of safety to ensure internal concerns or external pressures did not undermine planning
7. *motivation to learn* – experts recognised the importance of balanced debriefing and reflected on how their own decision making performance could improve.

Potential Disadvantages to WCS Thinking

Interviewees in Study 3 highlighted the importance of WCS thinking but identified a small number of potential disadvantages to WCS thinking. More specifically, interviewees identified three possible disadvantages:

1. fixation on extreme outcomes – decision makers may ignore the current situation and implement ineffective strategies and tactics (e.g., avoidance of aggressive direct attack, which is often the best way to minimise the size and destruction of a bushfire);
2. negative emotions – decision makers may become paralysed because of high anxiety levels associated with the imagined outcomes (e.g., avoidance of decisions);
3. overpredictions – decision makers may overestimate the possible future situations and disseminate inaccurate information (e.g., overestimate the possible rate of fire spread).

These potential disadvantages are consistent with findings of real-world decision making under stress, as explored further by McLennan et al. (Chapter 2, this volume). Regardless of these concerns, all interviewees reported that WCS must be considered in all bushfire decision making and highlighted that these concerns could be minimised through effective training. Experts also discussed methods for developing these WCS thinking skills (for example, scenario training).

Implications for Practitioners and Instructors

Recommendations for Australian Fire Agencies

The research findings described in this chapter emphasise the importance – and challenge – of adequately preparing for WCSs. The inconsistencies of WCS thinking reported by interviewees suggests further effort is required to ensure all decision makers adequately identify, plan and prepare for WCSs.

Practical recommendations have been developed for fire agencies based on the implications of my research findings. To improve WCS thinking, I suggest that practitioners can undertake the following actions:

1. recognise the importance of WCS thinking and adopt a range of strategies to improve the way WCS thinking is incorporated into decision making during bushfire incidents;
2. consider the extent to which a WCS decision tool could be developed to assist decision makers during incident management to better manage and prepare for possible WCS events;
3. develop and implement a comprehensive training regime to improve WCS thinking;
4. maximise the use of scenario exercises in training to improve performance of WCS thinking;
5. assess the efficacy of their current approach to mentoring and, if required, advance an evidence-based mentoring programme to ensure personnel develop and implement effective WCS thinking skills;

6. maximise the opportunities for learning and communicating lessons, both organisational and individual, to improve WCS thinking;

7. assess the extent to which the current practices of team composition effectively facilitate WCS thinking skills (e.g., preformed IMTs, as discussed in Hayes, Chapter 6, this volume);

8. examine the extent to which training decision makers in a formal process of critiquing plans (e.g., Klein's Premortem Method; Klein 2003) would improve WCS thinking (see also Brooks, Chapter 9 and Stack, Chapter 10, this volume);

9. consider the extent to which current planning protocols, upgrading practices, and associated training interfere with the ability of incident managers to perform effective WCS thinking.

These nine recommendations fit together to achieve an overall goal of enhancing WCS thinking and thereby improving decision making performance. A comprehensive improvement of WCS thinking skills requires a gradual agency-wide adoption of a range of strategies, including operational decision support tools and multi-faceted training (see recommendations 1, 2 and 3). Based on my research, training should incorporate the particular training approaches of scenarios, mentoring and formal processes of critiquing (see recommendations 4, 5 and 8). Furthermore, agency practices and standard procedures must encourage or support the consistent implementation of trained WCS thinking skills (see recommendations 7 and 9). Finally, developing a learning culture that ensures the organisation adapts and improves would support all the other elements required to improve WCS thinking (see recommendation 6).

Application of Recommendations

The recommendations refer to the development of a training regime and decision tool to improve WCS thinking, as well as describing the agency practices important for facilitating WCS thinking (for example, learning culture). Fire agencies should assess the implications of this research for their organisational practices to determine the nature and extent of response required to address each recommendation.

Some recommendations validate current fire agency practice, whereas others indicate areas where agencies could improve. Depending on the particular fire agency, these recommendations may be consistent with current agency practice and, therefore, provide evidence-based support for those practices. However, other recommendations will indicate areas where agencies require further work to ensure effective WCS thinking and improve decision making skills.

Efforts to improve WCS thinking should be carefully integrated into planning procedures for bushfire fighters, and associated training, to ensure efforts improve overall decision making in a cost-effective and practical manner. The recommendations target aspects of fire agency practice that will integrate within current structures and provide the most ongoing practical benefit in improving WCS thinking. In additional to improving WCS thinking, I expect that these recommendations are also likely to improve general decision making effectiveness and incident management outcomes.

Methods for Improving WCS Thinking

Based on the identified expert behaviours, some efficient methods were identified for developing WCS thinking skills and incorporating them into incident management practice. In particular, there were a range of methods relating to critiquing of decisions and plans, which appear to offer the greatest opportunities to improve WCS thinking (see also Brooks, Chapter 9, this volume).

Experts in Study 2 reported that discussing, checking and questioning plans were powerful techniques for minimising a number of barriers to WCS thinking, including underestimation, suboptimal attitudes, tunnel vision and situation characteristics. Critiquing and questioning are well-supported methods for improving decision making (LeSage, Dyar and Evans 2009). The ability to critique personal decisions is challenging because it requires strong metacognitive skills that allow decision makers to reflect on their own thoughts and feelings (Cohen, Freeman and Thompson 1995; McLennan et al. 2005). Numerous expert strategies implicated metacognitive skills of self-management and thought regulation, which previous research has identified as an essential element of effective real-world decision making (e.g., Cohen et al. 1995). Therefore, the importance of higher-order cognition for all decision makers cannot be overstated; developing metacognitive skills has the potential to improve WCS thinking by minimising various barriers (see also Frye and Wearing, Chapter 4, this volume).

Despite the importance of such behaviours, agency personnel in Study 3 reported that skills of critiquing needed stronger support in bushfire decision making and incident management. Similarly, an independent report into the Wangary fire in South Australia in 2005 recommended that standard procedures should 'build in authoritative "devil's advocate" processes...to ask the "what if" questions for bushfire[s] which have the potential to expand outside acceptable outcomes' (Smith 2005, p. 70). Critiquing should ideally be a normal practice for all decision makers because of the wide range of associated benefits, including reducing barriers to WCS thinking.

The Pre-mortem Method

One particularly effective and innovative critiquing method is the pre-mortem method developed by Gary Klein (2003), which facilitates the identification of possible WCSs and addresses many potential WCS barriers. The method requires group members to imagine a future scenario where the primary plan has failed and identify how the situation could have deteriorated to the point of catastrophic failure.

A pre-mortem is the hypothetical opposite of a post-mortem because a pre-mortem attempts to diagnose what might cause a catastrophic failure (i.e., the patient's death) before it happens. Instead of waiting until the catastrophic outcome occurs to identify the contributing factors, it attempts to investigate what might go wrong prior to them from developing into bad outcomes. The pre-mortem process is a straightforward tool for identifying worst case scenarios and strengthening plans to manage these possible situations.

The original pre-mortem process is usually carried out in a group of 8–12 and has six steps (Klein 2003):

1. Preparation – The facilitator introduces team members to the concept and describes the current situation or plan.
2. Imagine a fiasco – The facilitator asks team members to imagine looking into a crystal ball and seeing that the management of the incident has been totally unsuccessful. Rather than a simple failure, this failure is an embarrassing, devastating and catastrophic failure. There has been loss of life, loss of infrastructure, and loss of property. Unfortunately, the crystal ball cannot explain the reasons for this failure so team members need to find out: 'What could have caused this?'
3. Generate reasons for failure – Team members individually write down all the reasons they believe contributed to the catastrophic failure. The list of reasons developed by each team member may differ because it reflects their individual set of past experiences.
4. Consolidate the lists – The facilitator asks each team member to describe one reason for failure, which is recorded onto a master list. This continues until each team member has described every item on their list. The final master list provides a comprehensive record of the group's concerns.
5. Revisit the plan – The group choose the issues of greatest concern and discuss possible actions they could take to minimise the influence of these issues on outcomes. Other issues could be flagged for discussion at another time or in a different forum.
6. Periodically review the list – The consolidated list provides a reference of concerns that team members can review as the incident or project progresses. Reviewing these issues can re-sensitise the team to the problems that may emerge.

The pre-mortem relies on crystal ball thinking (i.e., an imaginary crystal ball acts as a perfect intelligence source; Cohen et al. 1995; Cohen, Freeman and Thompson 1997); the use of this technique in training has been shown to improve metacognitive processes. Behaviours encouraged by the pre-mortem process are also consistent with HRO behaviours (i.e., preoccupation with failure) (Weick and Sutcliffe 2007) and the expert strategy to manage underestimation by requiring decision makers to 'assume the worst and plan accordingly' (see Table 3.1).

In addition to minimising risk underestimation, the pre-mortem has the potential to strengthen plans, highlight assumptions, identify unintended consequences, expand timeframes considered, reduce tunnel vision and minimise optimism bias. Without using the term WCS, the pre-mortem leads decision makers to directly identify possible WCSs and actions to manage those possibilities. Therefore, this technique could also avoid the anxiety and fixation that a small proportion of decision makers may experience as a negative side-effect of WCS thinking.

Group evaluation, as required by the pre-mortem, strengthens plans because respectful dissent and open communication can isolate assumptions and biases, which improves WCS thinking. Group critiquing requires a supportive team culture that encourages others to voice opinions, raise concerns and question processes. Such positive interpersonal communication is a crucial component of successful team functioning and incident management effectiveness (Flin 1996; McCann and Pigeau 2000; Okray and Lubnau II 2004; see also Owen, Chapter 7, this volume). Thus, the pre-mortem has the potential to reduce between-group tensions and other interpersonal issues by encouraging effective communication.

The steps of a typical pre-mortem may be adapted to suit different purposes. For example, an individual pre-mortem could be used in training to develop the brainstorming and anticipation skills of a potential Incident Controller. After extensive use in training, decision makers may automatically employ the processes behind the pre-mortem of critiquing decisions and plans during incidents. Therefore, in addition, a pre-mortem can be used during an incident as a decision making tool to check decision making and planning. If the pre-mortem proved to be a valuable decision tool in improving WCS thinking, a concise version could be incorporated into the standard procedures of incident management team functioning to facilitate group discussion and critique of plans. However, the integration of such a technique into training or operational decision making must be done carefully, with an acknowledgement of the practical limitations of the work environment and the challenges inherent to incident management. Various Australian fire agencies have found this a useful approach in training and during operations to improve bushfire incident management.

The pre-mortem method, with careful development and implementation, appears to have potential as a useful tool for improving WCS thinking. Given the similarities between bushfire incident management and the planning procedures required in numerous other emergency management domains, the pre-mortem method could also be important for improving the management of floods, cyclones, tsunamis, and any other situation that requires robust and incisive planning procedures.

Implications Beyond Fire

The findings and recommendations of this research reflect possible improvements required to facilitate effective WCS thinking in a range of emergency service organisations (for example, police, flood management or urban fire agencies), as well as possible applications to numerous other real-world domains which share characteristics with the bushfire domain (for example, military command and control, and aviation). Although some differences exist between fire agencies and other domains, the many similarities result in the expectation that the practical research implications will resonate for personnel working in other domains.

Given that WCS thinking can be challenging for experienced incident managers, it is likely that community members may also find anticipating extreme events problematic. I was involved in post-incident research carried out by a Bushfire CRC taskforce after the February 2009 Black Saturday fires in Victoria. As part of the taskforce, I interviewed many community members affected by the fires.

Although WCS thinking was not the focus of the taskforce research, it was clear from some of the interviews that, prior to the fires, many people were unable to imagine how the situation could quickly deteriorate into disaster. Many survivors reported difficulties in planning for extreme outcomes and developing back-up options. Therefore, an important application of these concepts of WCSs may be in community decision making. Further research could investigate factors that might prevent community members from preparing for WCSs and provide guidance for developing new ways to improve WCS thinking.

The research outcomes also emphasise the value of reviewing past bushfire incidents. Every fire incident provides a substantial resource for organisational learning. All agencies should be identifying and exploring the successful decision making displayed by their personnel, as well as the mistakes that are inevitably made. The interview techniques used in this research (HFIP, WCSIP) could be valuable for fire agencies interested in performing robust incident reviews.

Conclusion

The purpose of this research was to explore the role of WCSs in real-world decision making. The integrated findings of all three studies provided insight into the potential barriers to WCS thinking, suggested potential methods for improving WCS thinking, and led to the development of practical recommendations for fire agencies.

The practical implications of this research indicated potential areas for improvement in fire agency practice and identified methods that will develop WCS thinking in a range of emergency service organisations (for example, police, flood management or urban fire agencies) and other real-world domains (for example, military command and control, and aviation).

Overall, research findings: (a) extended the theoretical understanding of WCS thinking in real-world decision making, and (b) led to practical recommendations for appropriately enabling WCS thinking.

Acknowledgements

The research was supported funding through the Bushfire Cooperative Research Centre. However, the views expressed are those of the author and do not necessarily reflect the views of the Board of the funding agency.

References

Adams, J. and Adams, P. (1961). Realism of confidence judgments. *Psychological Review,* 68(1), pp. 33–45.

Alexander, M.E., Mutch, R.W. and Davies, K.M. (2007). Wildland fires: Dangers and survival. In P.S. Auerbach (ed.), *Wilderness Medicine* (5th ed., pp. 286–335). Philadelphia, PA: Mosby.

Arkes, H.R. (1991). Costs and benefits of judgment errors: Implications for debiasing. *Psychological Bulletin,* 110, pp. 486–498.

Armor, D.A. and Taylor, S.E. (2002). When predictions fail: The dilemma of unrealistic optimism. In T. Gilovich, D. Griffen and D. Kahneman (eds), *Heuristics and biases: The psychology of intuitive judgment* (pp. 334–347). Cambridge University Press.

Beach, L.R. and Lipshitz, R. (1993). Why classical decision theory is an inappropriate standard for evaluating and aiding most human decision making. In G.A. Klein, J. Orasanu, R. Calderwood and C.E. Zsambok (eds), *Decision making in action: Models and methods* (pp. 21–35). Norwood, Ablex.

Boehm, G. and Brun, W. (2007). Intuition and affect in risk perception and decision making. *Judgment and Decision Making,* 3(1), pp. 1–4.

Buehler, R., Griffen, D. and Ross, M. (1994). Exploring the 'planning fallacy': Why people underestimate their task completion times. *Journal of Personality and Social Psychology,* 67, pp. 366–381.

Clark, R.C. (2008). *Building expertise: Cognitive methods for training and performance improvement* (3rd ed.). San Francisco, Pfeiffer.

Clarke, L. (2006). *Worst cases: Terror and catastrophe in the popular imagination.* University of Chicago Press.

Cohen, M.S., Freeman, J.T. and Thompson, B.B. (1995). *Training the naturalistic decision maker* (Tech. Rep. under contract MDA903-92-C-0053 for the Army Research Institute). Arlington, VA: Cognitive Technologies.

Cohen, M.S., Freeman, J.T. and Thompson, B.B. (1997). Training the naturalistic decision maker. In C.E. Zsambok and G.A. Klein (eds), *Naturalistic decision making.* Mahwah, Lawrence Erlbaum.

Crandall, B., Klein, G.A. and Hoffman, R.R. (2006). *Working minds: A practitioner's guide to cognitive task analysis.* Cambridge, MA: MIT Press.

Desmond, M. (2007). *On the fireline: Living and dying with wildland firefighters.* University of Chicago Press.

Fahey, L. and Randall, R.M. (1998). What is scenario learning? In L. Fahey and R.M. Randall (eds), *Learning from the future: Competitive foresight scenarios* (pp. 3–21). New York, Wiley.

Feltovich, P.J., Prietula, M.J. and Ericsson, K.A. (2006). Studies of expertise from psychological perspectives. In K.A. Ericsson, N. Charness, P.J. Feltovich and R.R. Hoffman (eds), *The Cambridge handbook of expertise and expert performance* (pp. 41–67). New York, Cambridge University Press.

Finucane, M.L., Alhakami, et al. (2000). The affect heuristic in judgments of risks and benefits. *Journal of Behavioral Decision Making*, 13(1), pp. 1–17.

Fischhoff, B., Slovic, P. and Lichtenstein, S. (1977). Knowing with certainty: The appropriateness of extreme confidence. *Journal of Experimental Psychology: Human Perception and Performance*, 3(4), pp. 552–564.

Flin, R. (1996). *Sitting in the hot seat: Leaders and teams for critical incident management*. Chichester, UK: Wiley.

Frensch, P.A. (1994). Composition during serial learning: A serial position effect. *Journal of Experimental Psychology: Learning, Memory, and Cognition, 20*(2), pp. 423–442.

Gigerenzer, G. (2007). Fast and frugal heuristics: The tools of bounded rationality. In D.J. Koehler and N. Harvey (eds), *Blackwell handbook of judgment and decision making* (pp. 62–88). Malden, Blackwell.

Gigerenzer, G. and Todd, P.M. (1999). Fast and frugal heuristics: The adaptive toolbox. In G. Gigerenzer, P.M. Todd and The ABC Research Group (eds), *Simple heuristics that make us smart* (pp. 3–34). New York, Oxford University Press.

Hoffrage, U. (2004). Overconfidence. In R.F. Pohl (ed.), *Cognitive illusions: A handbook on fallacies and biases in thinking, judgement and memory* (pp. 235–254). New York, Psychology Press.

Johnson, C. (2005). *A field and laboratory study of decision making by bushfire fighters* (Unpublished post-graduate diploma thesis). La Trobe University, Melbourne, Australia.

Johnson, C. (2011). *The Role of Worst Case Scenarios in Real-World Decision Making: A Study of Bushfire Fighting* (Doctoral dissertation). La Trobe University, Melbourne: Australia. Retrieved from National Library of Australia website: http://trove.nla.gov.au/version/201465165.

Johnson, C., Cumming, G. and Omodei, M.M. (2007a). The use of worst case scenarios in decision making by bushfire fighters. In K. Mosier and U. Fischer (eds), *Proceedings of the 8th International Naturalistic Decision Making Conference* (pp. 143–148). Retrieved from 8th International Naturalistic Decision Making Conference website: http://bss.sfsu.edu/kmosier/NDM8_Proceedings.pdf.

Johnson, C., Cumming, G. and Omodei, M.M. (2007b). The use of worst case scenarios in decision making to control wildfires: A field investigation. In M. Dollard, T. Winefield, M. Tuckey and P. Winwood (eds), *Proceedings of the 7th Industrial and Organisational Psychology Conference/1st Asia Pacific*

Congress on Work and Organisational Psychology (pp. 149–154). Melbourne, VIC: Australian Psychological Society.

Johnson, C., Cumming, G. and Omodei, M.M. (2008, September). *How worst case scenarios are considered by bushfire fighters: An interview study.* Paper presented at the Australasian Fire and Emergency Service Authorities Council Conference/Bushfire CRC International Fire Research Conference, Adelaide, SA.

Kahneman, D. and Klein, G.. (2009). Conditions for intuitive expertise: A failure to disagree. *American Psychologist,* 64(6), pp. 515–526.

Kahneman, D. and Tversky, A. (1982). The simulation heuristic. In D. Kahneman, P. Slovic and A. Tversky (eds), *Judgment under uncertainty: Heuristics and biases* (pp. 201–208). Cambridge University Press.

Keller, P. (ed.). (2004). *Managing the unexpected in prescribed fire and fire use operations: A workshop on High Reliability Organizations* (General Tech. Rep. RMRS-GTR-137 ed.). Retrieved from Wildland Fire Lessons Learned Center Website: http://www.wildfirelessons.net/documents/MTU_Santa_Fe_ Workshop_rmrs_gtr137.pdf.

Klein, G.A. (1997). Developing expertise in decision making. *Thinking & Reasoning,* 3(4), pp. 337–352.

Klein, G.A. (1998). *Sources of power: How people make decisions.* Cambridge, MIT Press.

Klein, G.A. (2003). *The power of intuition: How to use your gut feelings to make better decisions at work.* New York, Currency/Double Day.

Koriat, A., Lichtenstein, S. and Fischoff, B. (1980). Reasons for confidence. *Journal of Experimental Psychology: Human Learning and Memory,* 6(2), pp. 107–118.

Larrick, R.P. (2007). Debiasing. In D.J. Koehler and N. Harvey (eds), *Blackwell handbook of judgment and decision making* (pp. 316–337). Malden, Blackwell.

LeSage, P., Dyar, J.T. and Evans, B. (2009). *Crew resource management: Principles and practice.* Sudbury, Jones and Bartlett.

McCann, C. and Pigeau, R. (eds). (2000). *The human in command: Exploring the modern military experience.* New York, NY: Kluwer Academic.

McLennan, J., Omodei, M. et al. (2007). Human information processing aspects of effective emergency incident management decision making. In M. Cook, J. Noyes and Y. Masakowski (eds), *Decision making in complex environments* (pp. 143–151). Aldershot, Ashgate.

McLennan, J., Pavlou, O. and Omodei, M.M. (2005). Cognitive control processes discriminate between better versus poorer performance by fire ground commanders. In H. Montgomery, R. Lipshitz and B. Brehmer (eds), *How professionals make decisions* (pp. 209–221). Mahwah, Lawrence Erlbaum.

McLeod, R. (2003). *Inquiry into the operational response to the January 2003 bushfires in the ACT* (Report No. 03/0537). Retrieved from the ACT Government website: http://www.cmd.act.gov.au/functions/publications/archived/mcleod_ inquiry.

Milkman, K.L., Chugh, D. and Bazerman, M.H. (2009). How can decision making be improved? *Perspectives on Psychological Science, 4*(4), pp. 379–383.

Okray, R. and Lubnau II, T. (2004). *Crew resource management for the fire service.* Tulsa, OK: PenWell.

Omodei, M.M., McLennan, J. and Reynolds, C. (2005). *Identifying the causes of unsafe firefighting decisions: A human factors interview protocol* (Tech. Rep. No. 1). Retrieved from Bushfire CRC website: http://www.bushfirecrc. com/downloads/D2_3%20Report%201_2005%20Human%20Factors%20 Interview%20Protocol.pdf.

Orasanu, J. and Fischer, U. (1997). Finding decisions in natural environments: The view from the cockpit. In Zsambok and G.A. Klein (eds), *Naturalistic decision making* (pp. 343–357). Mahwah, Lawrence Erlbaum.

Over, D. (2007). Rationality and the normative/descriptive distinction. In D.J. Koehler and N. Harvey (eds), *Blackwell handbook of judgment and decision making* (pp. 3–18). Malden, Blackwell.

Pliske, R.M., McCloskey, M.J. and Klein, G.A. (2001). Decision skills training: Facilitating learning from experience. In E. Salas and G.A. Klein (eds), *Linking expertise and naturalistic decision making* (pp. 37–53). Mahwah, Lawrence Erlbaum.

Proctor, R.W. and Vu, K.-P.L. (2006). Laboratory studies of training, skill acquisition, and retention of performance. In K.A. Ericsson, N. Charness, R.R. Hoffman and P.J. Feltovich (eds), *The Cambridge handbook of expertise and expert performance* (pp. 265–286). New York, Cambridge University Press.

Russo, J.E. and Schoemaker, P.J.H. (1992). Managing overconfidence. *Sloan Management Review*, 33(2), pp. 7–17.

Slovic, P., Finucane, M.L. et al. (2004). Risk as analysis and risk as feelings: Some thoughts about affect, reason, risk, and rationality. *Risk Analysis*, 24(2), pp. 311–322.

Smith, B. (2005). *Report of independent review of circumstances surrounding eyre peninsula bushfire of 10th and 11th January 2005 (Wangary Bushfire).* Retrieved from the NSW Rural Fire Service Library website: http://publiclibrary.rfs.nsw. gov.au/CommonDocuments/ReviewEyrePeninsulaBushfire.pdf.

Sunstein, C.R. (2007). *Worst-case scenarios.* Cambridge, Harvard University Press.

Tversky, A. and Kahneman, D. (1973). Availability: A heuristic for judging frequency and probability. *Cognitive Psychology*, 5, pp. 207–232.

Tversky, A. and Kahneman, D. (1974). Judgement under uncertainty: Heuristics and biases. *Science*, 185, pp. 1124–1131.

Weick, K.E. (1995). Organizing and failures of imagination. *International Public Management Journal*, 8(3), pp. 425–438.

Weick, K.E. and Sutcliffe, K.M. (2007). *Managing the unexpected: Resilient performance in an age of uncertainty* (2nd ed.). San Francisco, Jossey-Bass.

Weinstein, N.D. (1980). Unrealistic optimism about future life events. *Journal of Personality and Social Psychology*, 39(5), pp. 806–820.

Chapter 4

What Were They Thinking?
A Model of Metacognition for
Bushfire Fighters

Dr Lisa M. Frye
School of Psychological Sciences, The University of Melbourne, Australia

Prof. Alexander J. Wearing
School of Psychological Sciences, The University of Melbourne, Australia

Introduction

Bushfires occur in Australia every year. Fortunately, most bushfires are quickly contained by firefighting crews and therefore remain small (e.g., see report by Fire Services Commissioner of Victoria 2013). However, sometimes they become large-scale bushfires, particularly in adverse weather conditions. The worst large-scale bushfires have typically involved extensive fire spotting, which occurs when embers travel ahead of the main fire front to start new fires (e.g., see Teague, McLeod and Pascoe 2009, p. 46). Loss of life has also been associated with a sudden wind change, which rapidly turns the flanks of a bushfire into a new and much larger fire front (see Teague et al. p. 42). These, and many other conditions, mean that all bushfires involve some level of risk. It also means that large-scale bushfires can involve extreme time pressures, high stakes and data overload. Sometimes decisions are made in conditions where few things can be controlled or manipulated. In this respect, large-scale bushfires are an example of complex (or macrocognitive) decision environments (Klein et al. 2003, p. 81).

In these types of conditions people can't always make perfect decisions. Instead, they may need to make decision tradeoffs (Hoffman and Woods 2011). For example, considerable time pressures may require fireground commanders to make a tradeoff between efficiency and thoroughness (called bounded cognisance), or between managing risks themselves and delegating those responsibilities to others (called bounded effectiveness). In extreme conditions, they may also need to defer the goal of containing a bushfire in order to protect life and property (called bounded responsibility), or to choose between the safety of their own fire crews and the safety of other members of the community. These are difficult decisions, and fireground commanders describe feeling 'damned if you do, damned if you don't' in these situations (see Figure 4.2).

Working Under Pressure

Previous investigations have also shown that bushfire fighters may experience 'stress, fear, panic and a collapse of clear thinking' on the fireground (Putman 1995). This is not surprising, because bushfires are dangerous, and humans naturally engage a fight, flight or freeze response when they are threatened. Similarly, incident management teams may experience cognitive overload during extreme events (e.g., Teague, McLeod and Pascoe 2009, p. 235). Again, this is not surprising because there is a limit to how much information humans can process, particularly when they are under pressure (McLennan, Holgate and Wearing 2003; Miller 1956). The ability to regulate cognitive and emotional responses is therefore a core skill for people who deploy to large-scale bushfires.

Expertise

Fortunately, some people report feeling cognitively in control under these types of conditions and they manage to perform well most of the time. For example, researchers in Australia have found that superior firefighters felt cognitively in control during urban structural fires, while their poorer performing peers felt cognitively overloaded (McLennan, Pavlou and Omodei 2005). The researchers concluded that the superior firefighters had better self-regulation skills. Johnson (2011) also found that expert fireground commanders used self-management techniques during large-scale bushfires. For example, they reflected on their own thought processes and actively managed stress and cognitive overload. Effective incident management teams (IMTs) also use team metacognition to manage cognitive overload (McLennan et al. 2006). All of these studies suggest that self-awareness and self-regulation are important for individual and team performance during large-scale bushfires (see Figure 4.1).

There has also been a substantial amount of research conducted internationally about how people make decisions in dynamic, uncertain and time-pressured situations (like bushfires). For example, Klein, Calderwood and Clinton-Cirocco (1988) found that expert fireground commanders recognise cues in a new situation and quickly implement strategies that have worked for them in the past. This is called Recognition Primed Decision Making (see Bremner, Bearman and Lawson, Chapter 8, this volume; Klein 1999). Several researchers also find that heuristics, or rules of thumb, enable experienced practitioners to make decisions very quickly (Gigerenzer and Goldstein 1996). The decisions aren't always perfect, but they are often good enough to get the job done (Connolly 1999; Kahneman and Klein 2009). Experienced people are also able to critique and correct their own decisions on the job (Cohen, Freeman and Wolf 1996). This means that they don't always need to develop a perfect strategy, or make perfect decisions, in the first instance. Indeed, it may be highly inefficient for them to develop perfectly detailed plans at the outset, since those plans may quickly become redundant. Instead, their initial course of action can be refined as a situation evolves and as more information becomes available. In artificial intelligence this is called a metacognitive loop (Anderson et al. 2006). Furthermore, Anderson et al. proposed that a relatively small number of heuristics (rules of thumb) would be sufficient to enable people to adapt to most new situations. Heuristics and metacognitive skills therefore enable people to learn quickly, and to adapt to complex decision environments (like large-scale bushfires).

Metacognition

Metacognition refers to thinking about and managing our own thought processes (Cohen, Freeman and Wolf 1996; Flavell 1979). For example, writing a shopping list is a metacognitive activity. It shows that we understand the limitations of our own memory, and also that we can manage that limitation (by writing the list) to get the task done. Understanding the limitations of our own thinking is called self-awareness, and managing those limitations is called self-regulation. Figure 4.1 shows how these skills are applied during large-scale bushfires.

Self-awareness involves knowing your own strengths and weaknesses (Figure 4.1). For example, it involves knowing how you think when you are under pressure. It also involves knowing the limits of your own ability. In this respect, experience is important because few of us know the limits of our own thinking until we have been tested. Similarly, feedback from other people is important. It enables us to calibrate our own perceptions with the perceptions of those around us. Regular feedback and experiences are therefore critical for developing self-awareness.

(metacognitive knowledge)   (cognitive control)

Self-Awareness   Self-Regulation

*'It's a recognition of your **capability**, it's your own **self**...inner recognition of...this is what I'm capable of. It's just years of **experience**...but it's the **trigger** to say, **flick the switch**, no, this is going to go beyond me.'*

*'...It's **automatic**. Contingency this, contingency that, what happens if? If I just sit here now and talk about it I go WOW! But I just do it **automatically**. I don't even think about it, and people go – gee you're **pretty relaxed**.'*

Figure 4.1   **Examples of metacognition described by an expert fireground commander**

Once we know our own strengths, weaknesses and performance limits, we can then use that knowledge to manage our thinking (called self-regulation). This is particularly important for complex tasks, and also when there are significant time constraints. Self-regulation is important even when there are checklists and rule-based procedures to follow, because people still need to know 'when to do what' (e.g., Dorner 1990). Aviation provides a good example of how self-regulation and rule-based procedures are used together.

An Aviation Example

Valot (2003) used flight simulators to explore how experienced pilots keep their thinking on track while they are flying a plane. The pilots were first asked to prepare a flight plan for a particular route, and then to fly that route in the simulator. Valot found that the pilots did not use rule-based procedures to achieve a perfect flight. In fact, their planned flights never matched their actual flights. Instead, they used knowledge from their previous experiences (metacognitive knowledge) to keep the flights within acceptable tolerance limits, and to avoid the *dreaded flight* or *worst-case scenario*. In this respect, the pilots tolerated imprecision with some procedures, but not with others. They therefore understood their safe operating space, or margin of manoeuvre (see Woods and Branlat 2010).

The pilots in Valot's (2003) studies also described rules of thumb (heuristics) which enabled them to keep track of highly dynamic activity, and to manage risks, memory, the chronological distribution of tasks, and the distribution of cognitive load between themselves and other people (and automated systems). These findings are important, because they show that even in highly regulated environments, where there are rule-based and standard operating procedures, practitioners still need to regulate their own thinking (see also Adams and Ericsson 2000). That way, they can implement the right procedure at the right time. They can also avoid overcorrecting when it is not required, and coordinate their own efforts with the efforts of other people (i.e., manage the distribution of cognitive load). These are the same types of issues that people face during large-scale bushfires, and like Valot, we chose to explore them using simulations.

Method

The examples in this chapter are from three studies detailed in Frye and Wearing (2011). In these studies we used think-aloud protocols to gather qualitative data about the thoughts and behaviours of experienced career and volunteer bushfire fighters (as recommended by Pressley 2000). One study involved interviews with practitioners on the fireground (see Omodei, McLennan and Reynolds 2005), and two subsequent studies involved visual-cued recall debriefs after command post simulations of the same fire (Frye and Wearing 2011). The data therefore included maps, photos, aerial images, subjective experience questionnaires, expert observer ratings, and records of simulation trials and human factors interviews. We used a grounded theory approach to identify patterns and regularities in the data (also recommended by Pressley 2000), and those patterns were coded using the NVivo8 Qualitative Analysis software (QSR International 2008).

Career and Volunteer Bushfire Fighters

We used a small number of very detailed interviews for these original studies ($n = 4$ on the fireground, $n = 2$ in a repeated measures simulation experiment, and $n = 4$ in a high cognitive load simulation rated by expert observers). The studies had a high degree of ecological validity (Frye and Wearing 2011, p. 40) and the results are consistent with subsequent case studies involving bushfire fighters and residents. Ecological validity is the extent to which the findings of the research apply in the real world (versus the laboratory). In this case, the studies provided a rich source of qualitative data about how experienced career and volunteer bushfire fighters (more than 10 years' experience) made decisions during a large-scale bushfire. The command post simulation is now used for training Incident Controllers of a rural fire service.

Two participants in our simulation studies were rated particularly highly, and they were considered to be experts by experienced Level 3 Incident Controllers who observed the simulations. This means that we were able to compare the skills of these experts (who reported feeling mostly in control), with their peers (who described feeling frequently overloaded in high cognitive load conditions). In this chapter we use a model of metacognition to show how the expert fireground commanders kept their thinking on track, and thus avoided errors associated with cognitive overload. We also show how other fireground commanders became overloaded and susceptible to errors.

High Cognitive Load (Stress)

For this research, cognitive load referred to the amount of information that a person needed to process in order to perform their role. Our measures were subjective. For example, we were not interested in measuring the exact amount of information that was processed, as might happen in a laboratory experiment. Instead, we were interested in circumstances where fireground commanders felt that they were under pressure and susceptible to errors. In this respect, cognitive load was also a marker for stress.

Stress is a complex variable that affects almost every human endeavour. It involves physical, emotional and cognitive responses. For this research, we wanted to focus specifically on cognitive responses (and metacognition). Nonetheless, the fireground commanders' descriptions show that physical, emotional and cognitive responses are closely related during large-scale bushfires (Figure 4.2).

"**hurry up and wait**"   "well this could be very **interesting**"   "**Damned if we do,**
"farting around"   "very concerned"   **damned if we don't**"
"pretty routine"   "a very scary concept"   "desperately trying"
"**everything was**   "well, this could be very nasty"   "in a bit of a panic"
under control"   "it all got very **hairy** from there on"   "it all fell to **crap**"
"lulled you into a **false**   "that sort of rang alarm bells"   "a lot of panic"
sense of security"   "it was pretty **hairy**"   "adrenalin starts pumping"
                                        "jaws are dropping a bit"   "scared the **crap** out of him"
                                                                                           "screaming and yelling"
                                                                                           "it all went to shit in a big hurry"

**Figure 4.2   Examples of bushfire fighters' descriptions of cognitive load
(stress) on the fireground**

Fireground commanders rarely used the word *stress* during these interviews
(see Figure 4.2). Nevertheless, their descriptions paint a clear picture of the
decision environment. Figure 4.2 also shows that the firefighters' descriptions of
cognitive load were particularly influenced by time and motion issues, such as
the amount of time that was available to build up a picture of the situation (called
situational awareness), to make decisions, and to put those decisions into action
(e.g., 'pretty routine' versus 'in a bit of a panic'). We were most interested in
high cognitive load conditions. That is, situations where people feel 'damned if
they do, damned if they don't'. These are the situations where constraints and
decision conflicts become most apparent. They also lead to cognitive overload and
human errors.

Cognitive Overload

For example, in one of our high cognitive load simulations an experienced
fireground commander forgot to deploy a fire tanker to protect a bulldozer. He was
not oblivious to safety issues (or the standard operating procedure); in fact he was
so pre-occupied with the safety of one fire crew (facing a burnover situation) that
he lost sight of the big picture, and therefore the safety of another fire crew (also
facing a burnover situation):

> …so focussed on one (tanker) that I lost focus on the other, and I end up having
> to go through the same scenario again with them. Can you get out? Are you
> safe?…this is when I realised, Bugger!…I'd missed one of our cardinal rules…I

didn't have a fire unit with the dozer...I need to be stepping back again, to what's coming at them around the corner, but my big (problem) was the speed of the way things were happening at the incident...we should have faith in the crew leaders to (manage safety), but at the end of the day, in this position, you're responsible for that safety. (Quote 1)

These types of competing cognitive demands are common during large-scale bushfires. Indeed, these types of conditions occurred after our simulation experiments, during the 2009 Victorian Black Saturday Bushfires (see Teague, McLeod and Pascoe 2009). Standard operating procedures (SOPs) can help in these situations, but they do not necessarily improve safety (or productivity) unless practitioners can successfully implement them in real-world conditions. This means that procedures need to be concise, and be practised regularly enough to be implemented automatically. However, in this high cognitive load condition (Quote 1) the demands of implementing multiple procedures at once exceeded the cognitive ability of the fireground commander and he described feeling overloaded and making errors.

Similarly, additional firefighting resources may sometimes assist, but the effort involved in coordinating them can also increase cognitive load. In fact, participants in our studies rejected additional firefighting crews, both on the fireground and also in our command post simulation experiments. They felt that the work associated with increased communication and coordination activities would lead to cognitive overload (again), and that they were therefore better off working with what they already had (a resource poor situation). Consequently, the ability to delegate, and to manage individual and team performance within large command and control structures, is essential to operations and safety on the fireground.

Human Errors

The experienced fireground commanders in our studies particularly emphasised the need for cognitive flexibility. That is, the ability to move between a macro view (big picture) and a micro view (detailed) of different aspects of the situation. Experienced fireground commanders also described making the same types of cognitive errors (under high cognitive load conditions) that they observe with trained novices in the field, such as:

1. focusing on what is happening in front of them, but losing sight of the bigger picture (or vice versa)
2. focusing on what is happening right now, but losing sight of what might happen next (or vice versa)
3. focusing on situational awareness, but leaving it too late to make decisions (or rapidly making decisions with inadequate situational awareness)
4. persisting with a goal, but failing to change plans when the situation changes (or failing to establish goals and priorities altogether)

5. accepting responsibility, but micromanaging or failing to escalate issues to others (or deferring too many responsibilities to others)
6. focusing on safety, but focusing so much on one safety issue that they lose sight of another (or overlooking safety while pursuing other goals).

We see two categories of error here: fixation errors and fibrillation errors. Fixation errors represent a type of tunnel vision. In these cases, focusing (or fixating) on one aspect of the bushfire may result in a blind spot to other aspects of the situation. For example, fixating on what is happening right now may result in a blind spot to what could happen next. This means that people may fail to notice important (but sometimes subtle) changes, and they may not act quickly enough to make and implement new decisions.

On the other hand, fibrillation errors occur when people change the focus of their attention so frequently that they fail to make effective progress on any task. They can therefore expend considerable effort on activities that have little positive effect. For example, some fireground commanders described 'desperately trying to keep up' during large-scale bushfires, whereas experts prioritised, delegated and even abandoned some tasks. We therefore find that calibration (self-regulation) is a core skill for fireground commanders. Below, we illustrate 'what they were thinking' on the fireground, and during simulations of a large-scale bushfire (Frye and Wearing 2011).

What Were They Thinking?

Fireground commanders need to be able to regulate their thinking in different types of situations. That is, they need to know what to pay attention to at different points in time. In our studies (Frye and Wearing 2011), experienced fireground commanders described paying attention to the same aspects of the situation over and over again, as shown in Figure 4.3.

The fireground commanders in these studies described a repetitive and adaptive cycle of thinking (Figure 4.3). We were initially surprised, because fireground command involves a lot of standard operating procedures (SOPs). These are often linear processes. We therefore expected to hear more descriptions about checklists and rule-based procedures. Instead, the fireground commanders described working in a dynamic and fluid environment, being adaptive, and adjusting to changing conditions. In fact, they said that being 'locked in' to a strategy was a common source of errors (see Figure 4.3). For example, several fireground commanders suggested that less experienced people may adhere 'too strictly' to the objectives outlined in an Incident Action Plan (IAP). They said that this could be a trap, because the situation might have changed since those objectives were written. Experienced fireground commanders therefore tolerated imprecision in some procedures and objectives, but not in others. These are the same types of judgements made by the pilots in Valot's studies (2003).

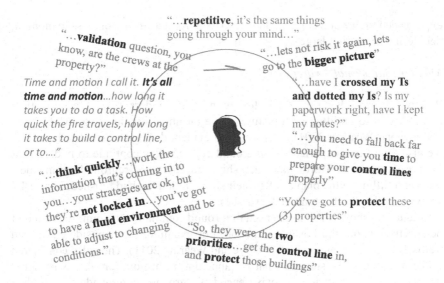

Figure 4.3   **A repetitive cycle of thinking, described by expert fireground commanders**

Decision Tradeoffs

Figure 4.3 also shows that some aspects of the fireground commander's decisions were in conflict. For example, they frequently juggled containment priorities (an offensive goal) with protection priorities (a defensive goal). These goals were complimentary when bushfires were small, because fires that are quickly contained pose fewer threats to life and property. However, these became competing goals once bushfires grew to be large-scale, particularly when resources were scarce. In these studies, there were significant differences between how experienced fireground commanders and expert fireground commanders handled this transition. We use a model of metacognition to show how fireground commanders managed their thinking in these 'damned if we do, damned if we don't' situations, including some of their decision tradeoffs (Hoffman and Woods 2011).

A Model of Metacognition

In artificial intelligence, researchers have suggested that intelligent systems (like people) use a metacognitive loop to notice changes in a situation, assess anomalies, and guide the system (or person) towards a solution (Anderson et al. 2006). That way, they can make adjustments to suit changing conditions and be resilient, or perturbation tolerant (see also Cohen, Freeman and Wolf 1996). In our studies, experienced fireground commanders described using a similar process to monitor, decide and act during large-scale bushfires.

Monitor (Situational Awareness)

Previous research has shown that fireground commanders allocate a lot of time and attention to building up a picture of the bushfire situation (e.g., McLennan, Omodei and Wearing 2001). This was also evident during our studies. However, we also found that experienced fireground commanders continue to update their picture of the situation throughout their shift. Indeed, some fireground commanders were still 'filling in the blanks' after their simulation trials had finished. Figure 4.4 shows how they built situational awareness.

Figure 4.4 shows that experienced fireground commanders monitored different perspectives about the bushfire situation. For example, they considered ground truths as well as the bigger picture (Frye and Wearing 2011). They also considered different timeframes, such as what is happening right now but also what could happen next. In particular, experienced fireground commanders described having a 'gut feel' for how to allocate their attention between these competing cognitive demands:

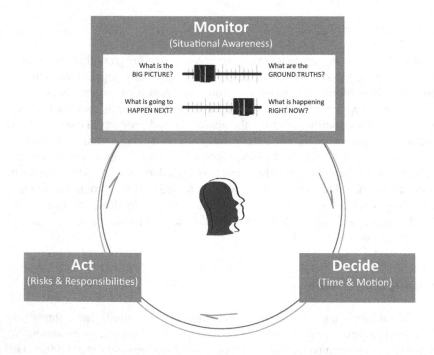

Figure 4.4   A metacognitive loop and decision tradeoffs for building situational awareness

...got to concentrate on what's in front of me, and work with what I've got, without being sort of reacting, but being proactive...I've got to take the limited amount of information that I've been presented...and try to fit that into a perception of the bigger picture. (Quote 2)

Like this fireground commander (Quote 2), all of the participants in our studies described 'mentally keeping a picture in their heads' of what was going on. In fact, one expert fireground commander described having a 3D spatial model in his head and he attributed that type of thinking to his role as an Air Attack Supervisor. This is consistent with previous research about situational awareness (Endsley 1995), meta-recognition (Cohen, Freeman and Wolf 1996), and also with a Recognition Primed Decision model (RPD; Klein 1999). The fireground commanders in these studies also described potential blind spots to situational awareness. For example, they said that less experienced people often struggle with monitoring the big picture and with predicting what might happen next:

...from what I saw, there was a lack of experience to deal with large task forces or strike teams, and realising how you need to keep them moving and keep them active and keep planning ahead. To my way of thinking, both divisions, east and west, were very much just keeping up as opposed to pre-planning. (Quote 3)

In contrast, experienced fireground commanders described examples from previous similar situations which enabled them to predict what might happen next. That is, they used previous experiences to conduct mental simulations. In this respect, our results support a Recognition Primed Decision model (see Cohen, Freeman and Wolf 1996; Klein 1999; Klein, Calderwood and Clinton-Cirocco 1988). This is important because it affects how quickly people can make decisions, and also whether they can recognise trigger points for changing goals during a large-scale bushfire.

Decide (Time and Motion)

For example, another concern for fireground commanders is whether to make decisions quickly, or to wait until they have a more complete picture of the situation instead. This can be described as an efficiency-thoroughness tradeoff (or bounded cognisance; Hoffman and Woods 2011). It recognises that in some contexts, decisions cannot be both perfectly timely and perfectly accurate (also described as a speed-accuracy tradeoff). For example, one fireground commander felt uncomfortable with the pace of his decision making during the simulation. He said that he felt compelled to act, and that in hindsight he would prefer to wait until he had more situational awareness to make decisions. Another fireground commander took the opposite view:

...if you make a good decision, (then) I don't have to do anything, and if you make a bad decision I (or someone else) can fix a bad decision. (But if) you make no decision at all (then) you kill someone. So I'd rather someone make a bad decision, and then we go 'bad mistake', and then it's corrected. But to stand out there, or in this scenario, and go 'um, um'...that's when you kill someone. (Quote 4)

According to this fireground commander (Quote 4), making no decision at all represents an error of omission (lost opportunity), and is more dangerous than making the wrong decision, or an error of commission (mistake). However, other studies have found that making hasty decisions can also compromise safety on the fireground (e.g., Bearman and Bremner 2013). For this reason, we recommend specific further investigation about the speed-accuracy tradeoff, and also deliberate practice for fireground commanders who make this type of judgement during large-scale bushfires (see Figure 4.5).

Most bushfires in Australia are quickly contained by firefighting crews and therefore remain small (e.g., see report by Fire Services Commissioner of Victoria 2013). This is usually achieved by immediately deploying resources to the fire to achieve a successful first attack. In this respect, containing the bushfire is an offensive goal.

When this approach is successful it also achieves the defensive goal of protecting lives and properties. However, bushfires can grow rapidly. This means that if a first attack is not successful, then fireground commanders need to distribute their cognitive resources (e.g., attention) and physical resources (e.g., fire tankers) between these two priorities:

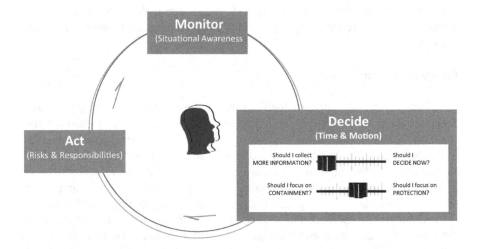

Figure 4.5   A metacognitive loop and decision tradeoffs for making timely decisions

...and I did this throughout. It was containment, property, containment, property...(Quote 5)

Fireground commanders can achieve this distribution of cognitive load when: fires are small, there are sufficient resources to deal with the situation, and they are experienced. However, even experienced fireground commanders felt overwhelmed during the high cognitive load and resource poor conditions in our simulations:

I actually start to focus so much on control (containing the fire)...that I lose sight of property protection...I'm still thinking that broad asset of (forest) that we want to protect, but not coming down to significant assets, like the structures. I've started to actually lose focus on them because I'm (still) thinking containment line...(Quote 6)

In contrast, we found that expert fireground commanders described clear trigger points for making a transition from containment goals to protection goals. Indeed, one expert described specific cues that enabled him to 'flick the switch' (see Figure 4.1). Another said that he realised as soon as he sized up the situation that there were not enough resources to contain the bushfire, so he concentrated on protecting life and property instead:

...it became apparent quite quickly that you're not going to stop this thing, but you've still got a whole lot of properties that you need to (protect). So, let's look at what we can do here...(Quote 7)

This expert was asked if he relied on experience to make these decisions (Quote 7). Initially he said no, that he looks at each situation on a case by case basis. However, a few minutes later he changed his mind and described previous similar situations where he had done the same thing. He also stipulated that he would switch back to containment goals when the conditions were right. He described those conditions in detail, and gave examples where he had changed goals like that in the past. In this respect, trigger points learned from previous experiences were used for making timely decisions about goals and priorities (consistent with the pilots in Valot's 2003 studies; and also with a Recognition Primed Decision model, Klein 1999; Cohen, Freeman and Wolf 1996).

Act (Risks and Responsibilities)

Large-scale bushfires can be dangerous, and they can involve large numbers of people (see Teague, McLeod and Pascoe 2009). For this reason, fireground commanders need to be able to work within a chain of command. That is, they need to be able to escalate some issues to more strategic positions, and delegate other issues to more tactical positions. This can be difficult for fireground commanders when they first step up to a new leadership role:

> …all of a sudden he had 12 trucks screaming and yelling at him what to do. And instead of letting the strike team leader run one section of it, he tried to run the whole lot himself…and when it all fell to crap, he couldn't keep up. (Quote 8)

One expert fireground commander said that delegating was 'a trust thing'. For example, he said that he has to be able to trust that the crew leaders are well trained. This is easier with people that he has worked with before, because he knows their capability. With new people, he said that he initially monitors them more closely (described as checking, monitoring, validating, getting ground truths). However, he also said that if he micromanages them, then he will lose sight of the bigger picture. This balance can be difficult, particularly when safety is involved (see Quote 1 and Figure 4.6).

Hoffman and Woods (2011) described five fundamental tradeoffs for making decisions in complex (or macrocognitive) work environments. In particular, they described safety as an example of a chronic goal which may be traded off against the acute goal of productivity (described as bounded responsibility). We believe that during large-scale bushfires, safety may involve a sixth tradeoff between two acute goals. That is, fireground commanders may need to make a tradeoff between the immediate safety of their own firefighting crews, and the immediate safety of other members of the community (see Figure 4.6). This is a particularly complex tradeoff.

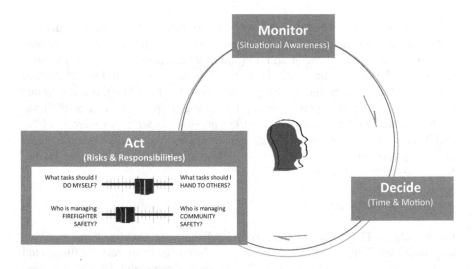

Figure 4.6   A metacognitive loop and decision tradeoffs for putting decisions into action

In our studies, bushfire fighters have always stipulated that their highest priority is to protect human life. Indeed, this forms the mission statement for several fire agencies in Australia. However, the decision to prioritise firefighter safety over other goals (such as protecting property) is described as a difficult one:

> ...the absolute focus on 'I will not lose an asset' could be the big undoing of somebody...in the bigger picture, you need to be able to say 'we're going to lose it'...that comes with experience, you know when it's going to happen, you deal with your feelings then, whether that be anger, frustration, you know, emotions, whatever. (Quote 9)

Firefighter safety is mandated in Australia by occupational health and safety legislation, and also the recommendations of previous judicial inquiries (e.g., Johnstone 2002). We therefore suggest that further specific investigation about this tradeoff is warranted. In particular, we recommend an exploration of how judicial processes and anticipatory regret (Connolly and Matthias 2011) influence this decision tradeoff during large-scale bushfires (or other emergencies). Such a study could highlight how fireground commanders make decisions about safety on the fireground.

Implications for Practitioners and Instructors

Bushfires are complex and so is human thinking. For this reason, no single conceptual model (or checklist) will perfectly explain what needs to be done when a small-scale bushfire turns into a larger one. However, in this chapter we have outlined the types of errors and decision tradeoffs that are described to us by experienced fireground commanders. In particular, we have illustrated a small number of metacognitive skills and heuristics which they use to regulate their thinking when they are under pressure. Our research shows that they have traditionally learned these skills on the job:

> Experience gives you the ability to read the dynamics and the changing situation. You predict what the fire's going to do, [based] on what you've been taught in the textbooks. These are the basic rules [and] principles of fire behaviour. Then you need to apply [them]. These are the experiences I've had in this type of country. This is what's worked, and this is what hasn't worked. So you learn from your mistakes obviously, and experience has a lot to do with it. (Quote 8)

This observation is consistent with a large body of empirical research (e.g., Ericsson 2006) and also with recommendations from crisis management experts (e.g., Leonard 2010; 't Hart 2010). Nonetheless, large-scale bushfires occur relatively infrequently and opportunities to gain appropriate workplace experiences may therefore be limited.

The Workforce

Bushfires are also fought in Australia by a largely part-time workforce. That is, the workforce comprises volunteers from country fire authorities in each state, and land managers from state government departments. This means that there are relatively few full-time bushfire fighters in rural Australia. Instead, people are deployed from their substantive roles to fight bushfires when they occur (see Frye 2012). This is different than in other complex work environments (such as aviation, medicine, or urban fire fighting), where practitioners frequently perform their role as a primary occupation, and therefore develop their skills in that context. Developing and maintaining expertise for large-scale bushfires is therefore a challenge. To meet this challenge, we recommend that fire agencies focus on two criteria for developing expertise, namely: learning in context, and learning calibration skills (or metacognitive skills, see Figure 4.1).

Learning in Context

In our studies, fireground commanders used previous experiences to build a picture of new situations. This included conducting mental simulations about what might happen next. They also described having a gut feel for the dynamics of the situation (e.g, time and motion issues), which enabled them to tolerate imprecision in some procedures, but not in others. In this respect, experienced fireground commanders understood their safe operating space (or margin of manouvre in aviation examples, see Adams and Ericsson 2000; Valot 2003; Woods and Branlat 2010). This was particularly important under high cognitive load conditions. For this reason, we recommend development activities that encourage fireground commanders to learn in context (Figure 4.7).

We also recommend that these activities (Figure 4.7) specifically address fireground commanders' ability to:

1. identify and correct blind spots to situational awareness (e.g., individually, and in teams)
2. conduct mental simulations (e.g., what might happen next and worst case scenarios)
3. regulate operational tempo (e.g., speed versus accuracy in decision making)
4. recognise trigger points for changing goals (e.g., from containment goals to protection goals)
5. manage risks and responsibilities within a chain of command (e.g., delegation and escalation)
6. apply safety procedures effectively (e.g., watchouts and safety drills).

Our research suggests that people stepping up to new leadership positions should also learn these skills in teams, so that they can calibrate their thinking.

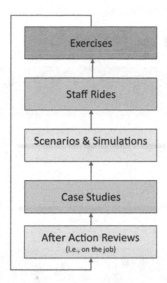

Figure 4.7   Examples of experiential learning and learning in context for large-scale bushfires

Learning Calibration Skills

For this research, we used think-aloud techniques to understand the metacognitive skills of experienced fireground commanders (see Pressley 2000). In particular, the visual-cued recall process provided us with an insight to 'what they were thinking' (see Frye and Wearing 2011). This included descriptions of errors and blind spots, as well as descriptions of self-regulation and expertise. We therefore propose that similar think-aloud techniques would enable emerging leaders (e.g., fireground commanders) to learn about metacognitive skills.

For example, we propose that fireground commanders can learn calibration skills from peers (i.e., buddy as coach), from teams (i.e., crew as coach), from supervisors (i.e., leader as coach), and also from mentors and subject matter experts. The model of metacognition presented in this chapter provides a starting point for those types of conversations (Figure 4.8).

This calibration check (Figure 4.8) could be conducted on the job, for example, at the beginning of a shift, during a tactical pause, and at the end of a shift. It could also be conducted during scenarios, simulations and exercises (see Figure 4.7). The intent is for emerging leaders to practise thinking in the way that expert fireground commanders do. Ideally, they would develop the self-regulation skills that enable experts to focus on the right things at the right time (i.e., keep their thinking on track), when they are under pressure.

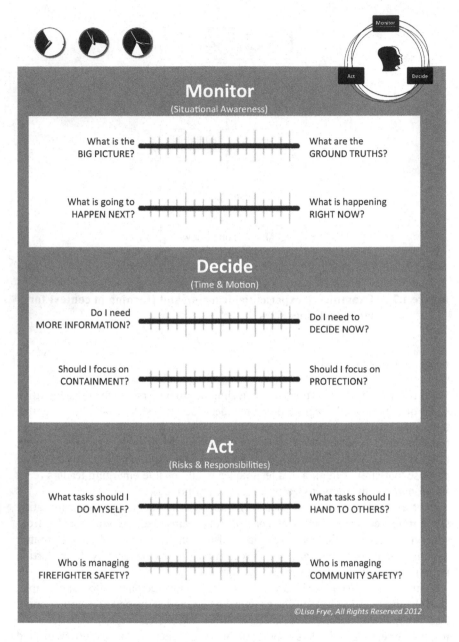

Figure 4.8   A calibration aid to support experiential learning

Conclusions

A large-scale bushfire is a complex decision environment. It involves many people, such as incident management teams, fireground commanders, firefighting crews and residents. Those people are also geographically dispersed. Each person must therefore monitor the situation (sometimes for hours or days), make a series of decisions (rather than one perfect decision), and coordinate their efforts with other bushfire responders. In this chapter, we have shown how experienced and expert fireground commanders use metacognitive skills to build up a picture of the situation (monitor), make decisions and put those decisions into action. We have also illustrated how they regulate their thinking, and thus avoid errors associated with cognitive overload, during high cognitive load conditions.

Bushfires also start small, but they can escalate quickly. This means that the transition from a small-scale to a large-scale bushfire can require a rapid change in thinking (like mentally changing gears). The current research shows that in these conditions people may not always be able to attend to everything that is going on (i.e., maintain situational awareness), accurately follow all of the requisite procedures and achieve all of the prescribed goals. They may therefore need to make decision tradeoffs (Hoffman and Woods 2011). In this chapter we have highlighted the types of decision tradeoffs that experienced fireground commanders make during large-scale bushfires. We also propose that there is a sixth decision tradeoff that applies to bushfire fighting, between the immediate safety of firefighting crews and the immediate safety of other members of the community. This tradeoff may be a particular example of bounded responsibility that applies when productivity goals are closely aligned to safety goals, as happens in emergency management.

Our studies suggest that fireground commanders use previous experiences (metacognitive knowledge) to regulate their thinking under high cognitive load conditions. In this respect, our findings are consistent with a large body of previous empirical research (e.g., Klein, Calderwood and Clinton-Cirocco 1988). In this chapter, we have also shown how think-aloud techniques provide insight to 'what the experts were thinking'. We suggest that similar techniques may be used for training fireground commanders about self-awareness and self-regulation during large-scale bushfires. In particular, we recommend learning and development activities that allow practitioners to address calibration skills, in context.

Further Research

The Australian Bushfire Cooperative Research Centre has collected a large number of fireground interviews in the last 10 years. For example, the Human Factors Interview Protocol (HFIP; Omodei, McLennan and Reynolds 2005) was used to conduct 120 interviews with bushfire fighters during the 2003–2006 Australian bushfire seasons. A Bushfire Research Taskforce also conducted 600 interviews in Australia with survivors of the 2009 Victorian Black Saturday bushfires

(Bushfire CRC 2010), and similar taskforces interviewed residents affected by the Perth Hills bushfires (2011), and the Tasmanian and New South Wales bushfires (2013). These interviews could be used to identify decision-making scenarios for use in experiential learning activities (e.g., case studies, simulations, staff rides or exercises). With this in mind, we have used this model of metacognition (Figure 4.4, Figure 4.5 and Figure 4.6) to analyse how survivors from a small rural community made decisions during the 2009 Victorian Black Saturday bushfires.

Acknowledgements

The studies referred to in this chapter were jointly funded by the New South Wales Rural Fire Service (NSW RFS) and the Defence Science and Technology Organisation (DSTO). However, the views expressed in this chapter are entirely those of the authors and do not necessarily reflect the views of these research sponsors. Richard Jones (of the firm One Eighty Seven & a Half) produced the graphic designs of this model of metacognition for our use in bushfire training.

Note

This chapter is based on a paper first published in the proceedings of the Australasian Fire and Emergency Services Authorities Council (AFAC) conference, Melbourne, 2–5 September 2013.

References

Adams, R. and Ericsson, A. (2000). Introduction to cognitive processes of expert pilots. *Human Performance in Extreme Environments*, 5(1), pp. 44–62.

Anderson, M., Oates, T. et al. (2006). The metacognitive loop I: Enhancing reinforcement learning with metacognitive monitoring and control for improved perturbation tolerance. *Journal of Experimental and Theoretical Artificial Intelligence*, 18(3), pp. 387–411.

Bearman, C. and Bremner, P. (2013). A day in the life of a volunteer incident commaner: Errors, pressures and mitigating strategies. *Applied Ergonomics*, 44, pp. 488–495.

Bushfire CRC. (2010). *Research task force: 2009 Black Saturday Bushfires*. Melbourne: Australian Bushfire Cooperative Research Centre. Retrieved on 12/11/2012 from http://www.bushfirecrc.com.au.

Cohen, M., Freeman, J. and Wolf, S. (1996). Metarecognition in time-stressed decision making: Recognizing, critiquing, and correcting. *Human Factors*, 38(2), pp. 206–214.

Connolly, T. (1999). Action as a fast and frugal heuristic. *Minds and Machines*, 9, pp. 479–496.

Connolly, T. and Matthias, J. (2011). Regret aversion in reason-based choice. *Research Collection Lee Kong Chian School of Business* (Open Access). Paper 3155. Retrieved from http://ink.library.smu.edu.sg/lkcsb_research/3155.

Dorner, D. (1990). The logic of failure. *Human Factors in Hazardous Situations*, 327(1241), pp. 463–473.

Endsley, M. (1995). Toward a theory of situation awareness in dynamic systems. *Human Factors*, 37(1), pp. 32–64.

Ericsson, K.A. (2006). The influence of experience and deliberate practice on the development of superior expert performance. In K.A. Ericsson, N. Charness, P.J. Feltovich, and R. R. Hoffman (eds), *The Cambridge handbook of expertise and expert performance* (pp. 39–68). Cambridge: Cambridge University Press.

Fadde, P.J. and Klein, G.A. (2010). Deliberate performance: Accelerating expertise in natural settings. *Performance Improvement*, 49(9), pp. 5–14.

Fire Services Commissioner of Victoria (2013). *Report on the 2012/2013 Victorian Bushfire Season*. Retrieved on 05/05/2013 from http://www.firecommissioner. vic.gov.au/latest-news/201213-season-overview/.

Frye, L. (2012). A multi-agency incident leadership framework. *Proceedings of the 3rd Human Dimensions in Wildland Fire Conference*, Seattle, 17–22 April 2012.

Frye, L. and Wearing, A. (2011). The Central Mountain fire project: Achieving cognitive control during bushfire response. *Journal of Cognitive Technology*, 16(2), pp. 33–44.

Flavell, J. (1979). Metacognition and cognitive monitoring: A new area of cognitive–developmental inquiry. *American Psychologist*, 34(10), pp. 906–911.

Gigerenzer, G. and Goldstein, D.G. (1996). Reasoning the fast and frugal way: Models of bounded rationality. *Psychological Review*, 103(4), pp. 650–669.

Hoffman, R.R. and Woods, D.D. (2011). Beyond Simon's slice: Five fundamental trade-offs that bound the performance of macrocognitive work systems. *IEEE Intelligent Systems*, 26(6), pp. 67–71.

Johnson, C. (2011). How bushfire fighters think about worst-case scenarios. *Firenote 77*, published by Bushfire Cooperative Research Centre, Melbourne.

Johnstone, G. (2002). *Report of the investigation and inquests into a wildfire and the deaths of five fire-fighters at Linton on 2 December 1998*. State Coroner's Office, Victoria.

Kahneman, D. and Klein, G. (2009). Conditions for intuitive expertise: A failure to disagree. *American Psychologist*, 64(6), pp. 515–526.

Klein, G. (1999). The recognition primed decision model. In *Sources of power: How people make decisions* (pp. 15–30). Cambridge, MA: The MIT Press.

Klein, G., Calderwood, R. and Clinton-Cirocco, A. (1988). *Rapid decision making on the fire ground*. Technical Report 796, US Army Research Institute for the Behavioural and Social Sciences.

Klein, G., Ross, K. et al. (2003). Macrocognition. *IEEE Intelligent Systems*, 18(3), pp. 81–85.

Leonard, H.B. (2010). Organizing response to extreme emergencies. Submission to the 2009 Victorian Bushfires Royal Commission. Doc. ID: EXP.3031.001.0018.

McLennan, J., Holgate, A. and Wearing, A. (2003). Human information processing aspects of effective emergency incident management decision making. *Proceedings of the Human Factors of Decision Making in Complex Systems Conference*, Dunblane, September 2003.

McLennan, J., Pavlou, O. and Omodie, M. (2005). Cognitive control processes discriminate between better versus poorer performance by fire ground commanders. In H.L. Montgomery, R., Brehmer, B. (eds), *How professionals make decisions* (pp. 209–221). Mahwah, NJ: Lawrence Erlbaum Associates.

McLennan, J., Holgate, A. et al. (2006). Decision making effectiveness in wildfire incident management teams. *Journal of Contingencies and Crisis Management*, 14(1), pp. 27–37.

Miller, G. (1956). The magical number seven, plus or minus two: Some limits on our capacity for information processing information. *The Psychological Review*, 63(2), pp. 81–97.

Omodei, M., McLennan, J. and Reynolds, C. (2005). Identifying the causes of unsafe firefighting decisions: A human factors interview protocol (Bushfire CRC Project D2.3 Safety in Decision Making and Behaviour, Tech. Rep. No 1). Melbourne: Australian Bushfire Cooperative Research Centre. Retrieved on 13/11/2012 from http://www.bushfirecrc.com.au.

Pressley, M. (2000). Development of grounded theories of complex cognitive processing: Exhaustive within- and between-study analysis of think-aloud data. In G. Schraw and J. Impara (eds), *Issues in the measurement of metacognition* (pp. 261–296). Lincoln: Buros Institute of Mental Measurements.

Putman, T. (1995). The collapse of decision making and organisational structure on Storm King Mountain. *Wildfire*, 4(2), pp. 40–45.

QSR International (2008). NVivo8 Qualitative Analysis Software. Melbourne: QSR International.

Simon, H.A. (1956). Rational choice and the structure of the environment. *Psychological Review*, 63, pp. 129–138.

't Hart, P. (2010). Organizing for effective emergency management. Submission to the 2009 Victorian Bushfires Royal Commission. Doc. ID: EXP.3031.001.0001.

Teague, B., McLeod, R. and Pascoe, S. (2009). *2009 Victorian Bushfires Royal Commission – Interim Report* (No. 225-Session 2006–09). Melbourne: Parliament of Victoria.

Valot (2003). An ecological approach to metacognitive regulation in the adult. In P. Chambers, M. Izaute, P. Marescaux (eds), *Metacognition: Process, function and use* (pp. 135–152). Boston, MA: Kluwer Academic Publisher.

Woods, D. and Branlat, M. (2010). Hollnagel's test: Being 'in control' of highly interdependent multi-layered networked systems. *Cognition, Technology & Work*, 12(2), pp. 95–101.

Chapter 5

The Role of Affect in Individual and Collective Performance in a Sociocultural Context

Dr Jan Douglas

University of Tasmania, Australia and Bushfire Cooperative Research Centre, Melbourne, Australia

Introduction

Many people who work in high-consequence environments enjoy the challenges and the complexities they face (Flach 1999). A number of research studies (e.g., Isen 2001; Isen and Reeve 2005) have shown that positive affect can influence people's work activity such as their quality of problem solving, decision making and caution in dangerous situations.

Yet, research shows that people working in high-consequence environments are also prone to negative affective experiences, which includes not having sufficient time to recover emotionally between critical incidents (Alexander and Klein 2001) and psychological distress (Burbeck et al. 2002). Moreover, McLennan et al. (Chapter 2, this volume) identified that stress can result in firefighters becoming too narrowly focused, taking longer to undertake tasks, having difficulties making decisions, and being more likely to make mistakes to the extent of impeding safety.

Both individual and collective affective experiences were important in this study. This is because firstly, individual affective experiences can influence motivation, cognition and performance; and secondly, because much of the work people do is carried out in groups and/or teams.

For the purpose of the study, affect was defined as the feelings that give rise to people's experiences (Barsade and Gibson 2007). Such feelings based on experiences, in turn, can influence cognition and motivation through efficacy beliefs. In addition, felt feelings (which are characterised as moods and emotions) can influence collective work activities and team climate.

The Role of Context

The position taken in the study was a sociocultural one where the 'biology/self and culture/society are considered inseparable' (Benton 1991, cited in Sturdy 2003, p. 90). Since what people experience happens in particular contexts, the sociocultural work environment is important. From this perspective affect is embedded in context. Context includes structures such as work groups and work roles that encourage particular interaction (Valsiner and van der Veer 2005 Owen 2001) and culture which is manifested in shared values and beliefs (Owen 2001). Both structures and cultures shape work experiences in particular ways. It will be shown in this chapter that these influence affect in incident management teamwork and have implications for performance.

Method

The study set out to explore the role of affect in fire incident management, and to examine the ways in which affect influences self-reported individual and collective performance. This raises the following questions:

1. What are the lived (felt) experiences of people who work in incident management teams?
2. How does affect intersect in a sociocultural context to influence incident management team performance?

Qualitative Inquiry

The core organising unit within the incident management framework is the incident management team, as discussed in Chapter 1 in this volume. Seventy semi-structured interviews were conducted across four states in Australia (see Table 5.1) with personnel who had various roles in incident management teams (see Table 5.2). Given that all personnel, with the exception of logistics officers, are required to have a firefighting background, and have the appropriate qualifications (e.g., incident controllers' training is at an Advanced Diploma level), the participants have many years of experience working in the industry and a minimum of eight years in incident management teams.

The interviews were conducted either face to face or over the telephone in the timeframe of 45–60 minutes. Participants were asked to talk about their experiences when engaged in incident management teamwork.

Table 5.1   Demographics: states and agencies

Tasmania		Victoria			New South Wales	Queensland
TFS	PARKS	MFB	CFA	DSE	NSWRFS	QFRS
10	9	8	12	10	9	12

Table 5.2   Demographics: participants' roles

Incident Controllers	Operations Officers	Planning Officers	Logistics Officers	Other (e.g., Situation Officers, Deputy ICs)
25	16	13	11	5

Data Analysis

All interviews were recorded, de-identified, transcribed and imported into NVIVO 8, which is a software program for qualitative analysis. Using a qualitative theory building approach the data was examined for people's affective experiences whilst engaged in work activity. In accordance with qualitative research the codes and categories were developed from the participants' responses and literature reviewed (Miles and Huberman 1984; Tesch 1990; Patton 2002; Creswell 2009). The data was analysed according to Tesch's (1990) system of de-contextualisation and re-contextualisation.

The Impact of Affect

The findings highlight the participants' affective states as they engage in incident management teamwork. They also illustrated how affect influenced their efficacy beliefs, which can assist in managing the demanding environment in which they work. The findings also discuss the way in which affect in a sociocultural context plays out and can lead teams to perceptions of either satisfactory, dysfunctional or optimal performance.

The Lived Experiences of Incident Management Personnel

When people are engaged in incident management they experience both positive (e.g., heightened self-efficacy and satisfaction) and negative affect (e.g., tension and frustration), which in turn can influence their performance.

Many participants, when talking about the challenges and complexities they faced, described the way they felt as experiencing 'joy' and 'pleasure', and 'feeling good in what we have done'. In the participants' narratives there was also a sense of achievement in the satisfaction of making progress in their tasks and being able to control what they referred to as a 'chaotic and almost out of control' situation. According to Brown (1998), people who have a sense of control over their situation gain a higher sense of self and are able to cope better with harsh conditions.

In managing pressure, a number of participants described the way they felt as 'being in a pressure cooker', while others said they felt 'drained'. Signs of physical, mental and emotional exhaustion were interwoven throughout the participants' stories. There was also evidence that people, at times, experienced a sense of vulnerability due to the amount of responsibility and the potentially high consequences of their decisions. This finding was congruent with the work of Brown and Brooks (2000) (undertaken in a related safety-critical domain), who found that the emotional climate of night nursing included feelings of fear and vulnerability because of the high responsibilities involved in working with limited resources. In my study, this pressure also led some participants to a turning point in how they felt about their work. For example, a number of participants indicated that in the past they enjoyed fire seasons (which in Australia are part of the landscape) and relish managing incidents. Now, however, they perceive their work as more of a problem that needs to be dealt with, which in turn is a source of stress. This, in part, is due to the community and media demands as well as the potential for litigation.

A number of incident management personnel talked about how the external pressure from the media, bureaucratic channels and the community at large has increased, with the community and governing bodies having higher expectations. Thus, part of their experience is to manage the media and community gaze. Managing the gaze has added higher demands and more complexity to managing incidents to the point where incident management teams become 'drained' and 'weary':

> If we were in a world where an incident management team only had to focus on the fire and where they were going to put people and the length of operational periods, you'd find our team would do it a bit easier because they only have to worry about the fire. [QFRS_10]

The following narrative provides an account of how an incident controller felt drained by the media. The quote also provides insights into the participant's sense of self and the relationship of his sense of self to the job. The Incident Controller was recalling a time when he was listening to radio talk-back on his way home after an incident. A listener phoned in to comment on what he thought about three dozers sitting on floats at the end of the road where he lives. The decision to remove three dozers (out of 25 dozers working on making a line around the fire) for 24 hours at that location was made by the Incident Controller as a part precautionary strategy. At this particular incident, it was necessary to undertake a major back burn to keep the fire under control and away from a pine plantation which was located at the rear of two sizable towns. If the dozers were not ready and in position when required it could take four or five hours to source them from the job they were doing to another location:

> The media attention in [year] really drained me...I'm driving home thinking 'Yep, we got the back burn in and it looks like [names of places] have been saved. We hadn't lost any houses.' (I turn on the radio) and this lady [radio announcer] is just asking for stories. This bloke rings in and says, 'This bloody

> mob don't know what they're doing. I suppose the operators are sitting around twiddling their thumbs'. So the announcer goes, 'Oh gee, that doesn't sound very good. Anyone else got any stories like this?' [DSE_06]

The Incident Controller continued his narrative about how angry and frustrated he felt with the public criticism of the decisions he had made. The participant noted that his anger and frustration still lingers on. Such emotions, if continually experienced, can lead to attitudinal exhaustion and disenchantment and ultimately result in burnout (Maslach and Jackson 1981).

Forgas and Williams (2002) contend most individuals 'have the strongest emotional reactions to information that touches on our sense of self' (p. 75). Embedded in the Incident Controller's story is evidence of how tensions within the self can reside (Markus and Nurius 1986). On the one hand, the Incident Controller felt good about his achievements (e.g., no houses were lost), which would enable him to experience the identity of a confident and competent incident controller. On the other hand, he has taken the comments he heard on the radio personally. Thus, his perceived social identity had changed (e.g., letting the community down rather than supporting the community).

There has been discussion in other chapters in this volume around the unforeseen emergent issues that fire emergency personnel have to deal with. The work personnel undertake is cognitively demanding. They need to consider multiple issues simultaneously and decisions are time-critical. How, then, do incident management personnel deal with such complexities? The findings of the study revealed that when engaged in work activity, personnel drew on both individual and collective efficacy beliefs. Thus, efficacy beliefs provide a coping mechanism to deal with the effects of a demanding environment (Williams et al. 2010).

The Role of Affect in Constructing Efficacy Beliefs

In managing the demands and complexities of the environment, the findings showed that personnel reported their need to anticipate and plan for worst case scenarios, as discussed by Johnson Chapter 3, this volume. As one participant said:

> I see it's going to be long [and] you plan for and I think being a number of steps in front of what the fire is doing makes a big difference ... and manage an incident with the ace of spades up your sleeve and try and keep three or four steps in front of it. [NSWRFS_02]

According to Bandura (1982, 1986), drawing on previous experiences (i.e., mastery experiences) is one of the most powerful sources of information to develop efficacy beliefs. The participant's comments also illustrate that incident management teams attempt to construct connections between past and present experiences as well as pre-empting and planning ahead. This supports the work of Tams (2006), who found that people's mastery experiences are constructed by building analogies between their present and past experiences.

Participants' also indicated that connecting past experiences with training is how they construct self-efficacy beliefs, as shown in the following quote, when an Incident Controller revealed how he prepares for unpredictable events:

> Initially you go 'Oh shit' and your stomach starts to churn a little bit and you go, 'What am I getting into? What am I facing?' and then I go into a zone and go, 'Okay, I've done this before [and] I've done the training. I've done it on the ground, I've done it all.' I go into a zone and I actually arrive quite relaxed. [DSE_01]

Whilst it is contended that this participant describes experiencing 'fear of the unknown', it is also suggests that this Incident Controller was able to control his emotion of fear because he experienced a sense of calmness and composure when he arrived on scene. Clearly, the Incident Controller had been successful in previously managing unpredictable events and reflecting on this raised his efficacy beliefs. He now believed (and had expectations) that he would be able to manage the incident efficiently and effectively. Again, this supports the work of Tams (2006), who found that when people are successful they develop confidence by focusing on the task and generalising from previous experiences.

The quote also suggests (as throughout other collected participants' stories) that there are linkages between experiencing a sense of 'flow' and self-efficacy beliefs. According to Csikszentmihalyi (1992), when a person is experiencing a state of 'flow', they have a heightened sense of self, and feel in effortless control. Flow is also about 'the way people describe their state of mind when consciousness is harmoniously ordered' (p. 6). It is suggested that when people are experiencing 'flow', their 'self-consciousness disappears' (p. 71) and they experience a heightened sense of efficacy which can lead to perceived 'higher levels of performance' (p. 74).

The findings identified that some personnel gained self-efficacy through vicariously observing team leaders (e.g., incident controllers and deputy incident controllers) and other team members who they believed were performing well and had self-confidence:

> I guess the feeling that you've actually made some progress. That's pretty satisfying it [the incident] was quite difficult and there was a period in the middle where we actually went from complete chaos, we [the incident management team] had a Deputy Incident Controller who kind of started pulling it into shape and we actually got another Incident Controller who was very good. In about three days the whole incident changed from being really a problem, and kind of difficult and out of control, to being in control, which we didn't think we'd do I think we were the only people who were able to hold anything that day. We had three fires and we kept them all under control and they [firefighters] got out,

like they escaped and we dealt with it really efficiently. So everyone was feeling
really good about that, you know the achievement. [NSWRFS_04]

As indicated in the narrative, the Incident Management Team had overcome
adversity and got on top of the fire, which engendered a sense of achievement. It is
argued that the participant (a planning officer) gained a heightened sense of self-
efficacy through observing and participating in a mastery collective performance
with other team members (i.e., vicarious experience – Bandura 1989), which in
turn assisted him in pursuing and reaching his goals. With the Planning Officer
comparing his team with other team performances, the quote suggests this provided
him with a mastery experience (Bandura 1997) and a good feeling of belonging to
a group that was somehow comparatively better than others (Brown 1998).

In summary, through exploring the lived (felt) experiences of personnel
who work in incident management teams, it is shown there is a tight connection
between feelings and performance. The results reported illustrate that affect, which
is embedded in incident management teamwork, permeates thinking and acting in
incident management teamwork activity, and has a role to play in constructing
efficacy beliefs.

From the Individual to the Collective in a Sociocultural Context

The question then remains, how is individual affect embedded in collective team
performance influenced? As Salas et al. (2007) stated, a team of experts does not
make an expert team. Simply bringing people together to perform together does
not create effective teamwork. This is because individuals within a team come
with their different experiences. For example, each team member has their own
roles and responsibilities. They also bring with them their different histories of
experiences of working in different organisations. Sometimes teams include
personnel from different organisations and individuals from these organisations
may come with different cultural values and objectives.

When individual experts come together they can be a team in name only, that
is, the team comprises of individuals who are differentiated by their tasks and
role responsibilities and who can at times end up working in 'silos' because some
individuals become so focused on their task. In Figure 5.1 these latent teams are
conceptualised as 'team differentiation'.

It is argued that how affect is managed collectively is the key to whether teams
remain latent (i.e., consisting only of differentiated individuals) or shift towards
dysfunctional (fragmented) or optimal (integrated) team performance.

The following findings will further illustrate the ways in which 'team
differentiation' can either lead to 'team fragmentation', and therefore the possibility
of dysfunctional collective outcomes, or 'team integration' where synergy is
present, thereby enabling teams to obtain optimal collective performance, as
shown in Figure 5.1.

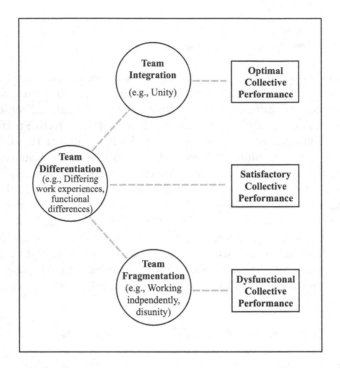

Figure 5.1   Differing teams and their performances

Affect and Differentiated Teams

More often than not, multiple agencies (e.g., fire agencies and land management agencies) are represented within the incident management team; thus, individuals can also be associated with their own agency identity as well as with the incident management team. When each function does not interact to the degree it should, tight boundaries around each function are formed. Incident management teams are also characterised by a number of functional work groups (i.e., control, operations, planning and logistics). See Chapter 1 in this volume for details of the different functions. Furthermore, functional work groups can create cultural variety within the incident management team because each function has its own set of norms, values and beliefs. Such differences can, at times, influence people's affective experiences and the way in which such groups undertake their work activity. When not managed well disconnects can occur and the team's full potential is not realised, as shown in Figure 5.1.

Ideally, all functional sections within an incident management team should collaborate with each other in order to meet subordinate goals. However, a number of participants talked about how, at times, these sections do not interact to the degree that is required for effective team performance. There needs to be consistent information flow within and between the Planning Section and Operation Section, for example, when making decisions about strategies and tactics. If each section is immersed in its own subgroup activities then functions become excluded and work is carried out in silos. A number of participants conveyed how 'frustrating' and 'unsatisfying' it is when they feel their section is working well together, yet they are unable to complete their task to the degree it should be because other sections do not pass on crucial information. One participant recalled a time when his planning section needed to produce an incident action plan and because he did not receive the required information he felt, 'it was a really crummy job [We produced] a piece of work we weren't particularly proud of you just feel like you're wasting time and you're not achieving what you could have achieved' [CFA_11]. Frustration can often lead to stress and have a negative influence on performance (McColl-Kennedy and Anderson 2002; Fineman 2003). It is argued that such individual emotions are shared and felt collectively (Roth 2007; Barsade and Gibson 2007) and may influence the team's performance.

When team members feel the work they have carried out is inadequate they are likely to experience a lowered sense of collective efficacy in collectively feeling they did not work collaboratively to produce a specific task (e.g., the incident action plan).

With such differences there is also potential for existing but not developed (i.e., latent) positive elements of team culture. It is when incident management teams move beyond their individual differentiation (to overcome silos) that they can create the synergy that leads to team integration (which will be discussed later in this chapter).

Groups (and teams) regulate their behaviour based on collectively held beliefs (Katz-Navon and Erez 2005). While there are limitations to working in silos, individuals focused on their individual tasks and responsibilities can still perform reasonably well (as illustrated in Figure 5.1). It is just that team members do not achieve the synergy that teamwork brings.

Affect and Fragmented Teams

The findings also showed that teams can experience fragmentation when affect is not managed well, and incident management teamwork is characterised by inconsistency, tensions and contradictions that are not successfully overcome. Team members' negative affective experiences included conflict that was not managed well. It was also found that when values and beliefs were not collectively held a lack of team orientation can occur, which leads to a negative affective experience based on individualistic behaviour at the expense of the team.

This, in turn, tends to impair team members' collective sense-making. Such negative affective experiences can lead to a team becoming dysfunctional. These are exacerbated by reported cultural differences in incident management teams. The data showed that in incident management teams there were times when cultural differences emerged through divisions based on 'who is perceived as old and new to the organization' (Keyton 2005, p. 59) or in this case histories of experience that are different in the team.

The following quote is from a deputy incident controller who talked about a time when differences within the team manifested in negative affect. At this particular incident the team changed midway through his first tour. The section leaders (i.e., the Operations Officer, the Planning Officer and the Logistics Officer) and some of the other senior team members left and were replaced with other personnel; thus, the team became comprised of 'old' and 'new' members. The 'new' team members had shared histories of experience because they had worked together before. The 'new' team members came in and assumed their senior management positions in an incident management team of 70 people:

> The dynamics [of the team] actually took, I reckon, 24 hours to settle down because they [new team members], without knowing it, alienated a lot of people [old team members]. I found it difficult because I'd been there for a few days and had the knowledge about where the incident was at, and yet I couldn't actually break in and have influence over this team 'cause they were such a close knit [sub] team.

> We had an incident controller who [was] a very, very competent incident controller, but maybe arrogant is the wrong word but [he was] very self-assured and very confident in the [new] team they'd brought in. So [he] had a high level of comfort with a small part of the team. [However], a lot of other people [on the incident management team] he didn't know and so [he] tended to fall back on the people that he knew…[CFA_02]

There is reference in the quote to the dynamics of the team changing when the shift changed and new members entered the team; thus, the dynamics of the team are influenced by the in-groups and out-groups within the incident management team. Again, this illustrates how subcultures can emerge through the way in which incident management work is organised based on the familiarity of some members. For further discussion on the role of familiarity see Hayes, Chapter 6, this volume. One of the elements that can negatively change in the dynamics of team is the level of trust amongst members. This is because, in part, trust is historically built through repeating interaction over time (Stokes, Lyons and Schneider 2011) and the degree to which cultural difference exists (Hughes et al. 2011).

Participants in this study also showed that incident management personnel place importance on having trust in the way in which people manage under uncertainty. When management teams become emotionally charged trust plays an important role. This is because it is important to know how people might express emotion and to what point that emotion might overpower their decision making and lead them, as one participant said, 'to the point they will snap' [NSWRFS_03]. Trust is defined by Mayer et al. (cited in Stokes et al. 2011) as 'the acceptance of risk or vulnerability from others based on positive expectations' (p. 13). It is suggested that limited or diminished trust in team members would be unfavourable amongst team members and could potentially impede on the team's functioning. Hayes, Chapter 5 in this volume, further expands on the linkages between familiarity and trust and its importance in teamwork.

The Deputy Incident Controller's story also illustrates there is an element of alienation (albeit without knowing) from the newly appointed team members towards the pre-existing team members. When working in an environment where alienation and disconnect occurs, it is likely team members' work activity is affected. Thus, it is suggested a lowered sense of mastery would be experienced (Bandura 1977), which can compromise managing complexities in time-critical environments. When this occurs there is a possibility that the team's collective performance becomes sub-optimal and even dysfunctional, as illustrated in Figure 5.1.

What is also of interest is that while the Incident Controller was seen as being competent and confident, it is inferred that this incident controller was not engaged with his whole team. It was not until the Deputy Incident Controller brought the situation to the Incident Controller's attention (i.e., successfully managing this tension and a potential source of conflict) that he became aware that he was not engaging the whole team. Creating such an inclusive environment is important because it can enable team communication which is a critical component in establishing safety culture (see Owen, Chapter 7, this volume).

Leadership style amongst a team can influence the extent to which a team member feels part of a team or not (Bass et al. 2003). If the Incident Controller in the story engaged with the whole team and concerned himself with how his team members felt (and were functioning) then the lasting negative feelings might not have developed. Moreover, leaders who effectively manage the emotional processes of their team have a positive influence on performance (George 2000).

Another way a team can become fragmented is when individualistic behaviour occurs at the expense of team orientation. Culturally, three forms of individualistic behaviour were found in the data representing negative stereotypes. According to Augoustinos, Walker and Donaghue (2006):

> Stereotypes are social representations: they are objectified cognitive and affective structures about social groups within society which are extensively shared and emerge and proliferate within particular social and political milieu of a given historical moment. They are socially and discursively constructed in the course of everyday communication. (p. 258)

Some participants (throughout a variety of agencies) talked about three individual negative stereotypes which they labelled as 'cowboys', 'shooting stars' and 'mavericks'. These negative stereotypes give us insights into behaviour that undermines effective teamwork.

There were a number of participants, in some organisations, who regarded some firefighters who work on the fireground as 'cowboys'. One participant described 'cowboys' as being 'very poorly disciplined who decided to, rather than follow the Incident Action Plan, they'd do their own thing' [QFRS_02], implying that rather than working with the Incident Management Team 'cowboys' work against them and work individualistically rather than with a collective orientation. The consequences of not adhering to the Incident Action Plan can put people's lives at risk. It is also undermining the Incident Controller's authority and goes against policies and procedures that are espoused in the organisational processes.

There are some firefighters who are less experienced than others and do not manage their own emotions. They were stereotypically regarded by some participants as 'shooting stars'. The following quote from an incident controller provided an example of the way in which 'shooting stars' can behave and the consequences of that behaviour:

> [On the fireground] there were too many shooting stars [all] over the place and it wouldn't have taken much [for] something bad to have happened... there's nothing more frustrating than being responsible for the safety of staff and not having control over it. They don't know because they are very small rural brigades that don't get big jobs frequently. [TFS_09]

The reference to 'shooting stars' in this respect indicates personnel who are not managing their emotions and are erratic, chaotic and potentially near panic. *The Australian Concise Oxford Dictionary (2002)* defines a shooting star as 'a small media moving rapidly and burning up on entering the earth's atmosphere'. In providing a cultural analysis of this characterisation, it could be argued that 'shooting stars' could be regarded as people who work and behave in an erratic and unpredictable manner. Such behaviour, from an individual, is not going along with the collective expectations successfully managing affect which includes being level-headed, having self-control and displaying calmness. When individuals behave erratically and are unpredictable it can potentially be quite disruptive and even dangerous. When an individual behaves in this way there is indication that his/her emotions are not being managed effectively. There was sense of frustration in the Incident Controller's story; he could also have been feeling a sense of fear for others' safety which inevitably would increase stress.

As discussed throughout the chapters in this volume, incident management work requires working under demanding conditions. When people are faced with stressful and dangerous situations, the body provides the natural stimulating chemical adrenalin which assists individuals to deal which such situations. The rush of adrenalin can give people a sensation of being fully alive. However, from some of the participants' stories, there are times when such experiences can foster dysfunctional performance, as illustrated in Figure 5.1. This is because for some people, the rush of adrenalin can make it difficult for them to adhere to structure and boundaries, which can once again lead to individualistic behaviour and thereby fragmenting the team effort. There were other personnel who were considered 'mavericks'; their behaviour indicates that their self-interest comes first.

This section has discussed that when beliefs are not shared, and some team members behave individualistically, there is an indication of a lack of team orientation. When there is lack of team orientation, the team tends to become fragmented. The next section discusses the way in which team integration can occur.

Team Integration and Managing Affect

From analysing the participants' narratives there is an indication that teams can become integrated when team members successfully overcome obstacles together and have a collective team orientation. This in turn enhances collective efficacy and performance.

The findings showed that when team members encourage and monitor their own and each other's affect and work collaboratively it can build a sense of collective confidence. The analogy of a football team can be applied to incident management teams – some teams work better than others:

> The football team that is running around without talking to each other on the football ground, they are playing footy. But the football team who are charged up, who are aware of each other, taking to each other are encouraging and supporting. Take that notion and put it into an incident management team. If you have got people who are supportive, encouraging and keeping the ball afloat, that is great...even though [people at times] are being short, direct and precise, they have still got a wink in their eye and they are saying 'Good on you, thanks very...can I give you a hand?' That in essence builds a team's confidence. [CFA_06]

Some participants felt a sense of integration when team members were 'bonding' and 'gelling' with each other indicating that the way in which team members function collaboratively together is an important element in teamwork. As one participant indicated, 'they [team members] might be good at what they do, but to be able to mesh together and work as a team, I think that's probably the key to it' [PARKS_9]. It is suggested that what this participant is describing is a sense of cohesion. Some studies (e.g., McDowell and Zhang 2009) have shown there is a link between team cohesiveness and good communication. When an incident management team has good communication they are more likely to acquire, interpret and control the flow of information so they are able to make sense of the environment and be prepared for any complexities and risks that might arise (see Owen, Chapter 7, this volume).

A sense of cohesion can also provide team members with the degree to which their affective needs (e.g., satisfaction) have been met (Carron et al., cited in Heuze et al. 2006). When team members feel satisfied or experience a heightened sense of collective efficacy they are more likely to make the correct decisions. This is because 'once emotions are evoked, they direct people's behaviour and can influence people's cognitive abilities' (Reus and Liu 2009, p. 253). It is important for team members to successfully manage affect because emotions can be transferred and felt collectively (Barsade and Gibson 2007).

Other participants described incident management teams as running 'smoothly', 'efficiently', and 'flowing' when they are working well. It is suggested that what people are collectively experiencing is a sense of fluidity or flow amongst the team. Given that 'flow' can assist people to pursue their goals and experience a heightened sense of self which can lead to higher levels of performance (Csikszentmihalyi 1992) (as mentioned earlier in this chapter), it is suggested that when people perceive others to have confidence in what they do and have the right skill sets it can contribute to a sense of 'flow' within the team. This indicates there are linkages between a heightened sense of collective efficacy and experiencing 'flow'. When team members are performing with a collective orientation and are successfully managing their own emotions and monitoring others there is a heightened sense of collective efficacy. Teams operating under these conditions are more likely to achieve their tasks and collective goals; they become more motivated and resilient to any difficulties. There is a linkage between performance and what Tasa, Taggar and Seijts (2006) called 'efficacy spirals'. That is, when people experience a heightened sense of collective efficacy, team members will more likely achieve optimal performance and collective goals. On the other hand, when people experience a lowered sense of collective efficacy, team members will be less motivated to perform well and are less likely to reach collective goals. Once again it is argued that successfully managing affect with the team is the key to a heightened sense of collective efficacy.

A number of participants talked about the importance of gaining a sense of collective confidence in the team's capability in order to overcome feelings of 'fear, pressure, vulnerability, being overstretched' and be able to work successfully in a team. Throughout participants' stories it was illustrated that collective confidence is experienced when team members have confidence in other members' knowledge, teamwork behaviours and problem-solving skills, and the significance they place on their team relationships to accomplish their desired outcomes. These elements then, become part of the shared histories of experience, which collectively supports the basis of the group's culture that underpins incident management teams.

Implications and Strategies for Practitioners and Instructors

While this study focused on one particular type of work activity, that is, managing wildfires, it is suggested that the implications of the findings and strategies that are recommended might be pertinent in other high-consequence domains.

Increased attention has been given to human factors in terms of individual and team decision making. This study has shown that people who work in incident management experience both positive and negative affect. This is because affect is embedded in individual and collective behaviour. The key to influencing the outcome of performance is how individuals monitor and manage their own affect as well as how they monitor, manage and mediate affect in others. If ignored this may not necessarily lead to negative performance, but it is unlikely to lead to an optimal one. If managed badly it is likely to lead to dysfunctional performance through team fragmentation. If managed well it can support team integration and optimal performance. Thus, it is of equal importance in professional development training that suitable attention is given to the role of affect on individual and collective decision making.

The findings also illustrated that when engaged in high-consequence work people can become physically and emotionally exhausted, which can influence their individual and collective performance. It is important, therefore, for all team members (including all leaders) to be trained to recognise indicators of stress and burnout, and be given strategies that will enable them to be manage stress so that stress levels will not increase to a point where they impact on team member performance and wellbeing. It is also important that members recognise the integrated part of affect in individual thinking as well as in team performance.

The data identified negative stereotypes of individuals who do not behave in a manner that supports collective orientation. Team leaders need to understand the role of affect and to be able to recognise the early warning signs if affect is not being managed well either individually or collectively.

Incident management requires a safety culture where attitudes, values and beliefs are shared to determine the degree to which all team members direct their attention and actions towards minimising risk to self and others. One of the negative stereotypes identified in the data and culturally labelled as 'shooting stars' were seen behaving in an erratic manner which is potentially dangerous when engaged in high-consequence activity. Thus, it is important for leaders to undertake professional development to equip them in managing difficult people. It would also be valuable for all personnel to receive professional development in emotional intelligence so that personnel gain a better understanding of themselves in terms of self-awareness and self-regulation and so they can learn to express negative emotions in a positive way to increase individual and collective performance. Emotional intelligence provides people with skills so they can:

1. identify emotions in themselves and others
2. use emotion to facilitate thought
3. understand the complexities of emotion and shifts in emotion
4. manage emotion in themselves and in others.

Conclusion

People working in such turbulent environments, such as those in incident management teams, need to manage a range of lived experiences, each with their requisite emotions. Indeed, when people are engaged in incident management work they experience both positive (e.g., heightened self-efficacy and satisfaction) and negative affect (e.g., tension and frustration) which needs to be managed. The chapter also discussed the way in which people construct their efficacy beliefs. To assist in dealing with managing pressure, incident management personnel can draw on self-efficacy and collective efficacy beliefs. The way in which individuals perceive their sense of self through social construction was also revealed.

From a sociocultural perspective there are different ways of situating individuals in their context. Affect, for example, can manifest in teams which can result in teamwork that is culturally differentiated, fragmented and integrated. All teams can be differentiated by characteristics of working in silos which has limitations, however, by large, team differentiation does not necessarily impede teamwork to any great extent; thus, overall teams can work to a satisfactory performance level. There are, however, times when such teams do not manage affect well and these can become fragmented. As a result, dysfunctional collective performance can transpire. Yet, teams with their differentiation can also use their communication skills to develop a collective orientation to monitor, manage and mediate affect both for individual and the collective of possible team synergy which can enable 'team integration'. Teams that are integrated can result in optimal collective performance.

Acknowledgements

The research was supported funding through the Bushfire Cooperative Research Centre. However, the views expressed are those of the author and do not necessarily reflect the views of the Board of the funding agency.

References

Alexander, D.A. and Klein, S. (2001). Ambulance personnel and critical incidents: Impact of accident and emergency work on mental health and emotional well-being. *British Journal of Psychiatry*, 178(Jan.), pp. 76–81.

Augoustinos, M., Walker, I. and Donaghue, N. (2006). *Social cognition: An integrated introduction.* London: Sage.

Bandura, A. (1977). Self-efficacy: Toward a unifying theory of behavioral change. *Psychological Review*, 84, pp. 191–215.

Bandura, A. (1982). Self-efficacy mechanism in human agency. *American Psychologist*, 37(2), pp. 122–147.

Bandura, A. (1986). *Social foundation of thought and action: A social cognitive theory.* Englewood Cliffs: Prentice Hall.

Bandura, A. (1989). Perceived efficacy in the exercise of personal agency. *The Psychologist*, pp. 411–421.

Bandura, A. (1997). *Self-efficacy: The exercise of control.* New York: W.H.Freeman and Company.

Barsade, S.G. and Gibson, D.E. (2007). Why does affect matter in organizations? *Academy of Management Perspectives*, 21(1) pp. 36–59.

Bass, B.M., Avolio, B.J. et al. (2003). Predicting unit performance by assessing transformational and transactional leadership. *Journal of Applied Psychology*, 88(2) pp. 207–218.

Brown, J.D. (1998). *The self*. McGraw-Hill.

Brown, R.B. and Brooks, I. (2002). Emotion at work: Identifying the emotional climate of night nursing. *Journal of Management in Medicine*, 16(5) pp. 327–344.

Burbeck, R., Coomber, S. et al. (2002). Occupational stress in consultants in accident and emergency medicine: A national survey of levels of stress at work. *Emergency Medicine Journal*, 19(3), pp. 234–8.

Creswell, J.W. (2009). *Research design: Qualitative, quantitative, and mixed methods approaches*. University of Nebraska, Lincoln: SAGE Publications, Inc.

Csikszentmihalyi, M. (1992). *Flow: The psychology of happiness*. USA: Harper & Row.

Fineman, S. (2003). *Understanding emotion at work*. London: Sage.

Flach, J. (1999). Beyond error: The language of coordination and stability. In E. Carterette and M. Friedman (eds), *Handbook of perception and cognition 2nd edition: human performance and ergonomics*. San Diego, London, Boston, New York: Academic Press.

Forgas, J.P. and Williams, K.D. (eds) (2002). *The social self: Cognitive, interpersonal, and group perspectives*. New York: Psychology Press.

George, J. (2000). Emotions and leadership: the role of emotional intelligence. *Human Relations*, 53, pp. 1027–55.

Heuze, J. P. Raimbault, N. et al. (2006). Relationships between cohesion, collective efficacy and performance in professional basketball teams: An examination of mediating effects. *Journal of Sport Sciences*, 24(1), pp. 59–68.

Hughes, S.C., McCoy, C.E.E. et al. (2011). Cultural influences on trust. In N.A. Stanton (ed.), *Trust in military teams*. Farnham: Ashgate Publishing Ltd.

Isen, A.M. (2001). An influence of positive affect on decision making in complex situations: Theoretical issues with practical implications. *Journal of Consumer Psychology*, 11(2), pp. 75–85.

Isen, A. and Reeve, J. (2005). The influence of positive affect on intrinsic and extrinsic motivation: Facilitating enjoyment of play, responsible work behavior, and self-control. *Motivation and Emotion*, 29(4), pp. 295–323.

Katz-navon, T.Y. and Erez, M. (2005). When collective efficacy and self-efficacy affect team performance. *Small Group Research*, 36(4), pp. 437–465.

Keyton, J. (2005). *Communication and organizational culture: A key to understanding work experiences*. Thousand Oaks, CA: Sage Publications.

Markus, H. and Nurius, P. (1986). Possible Selves. *American Psychologist [PsycARTICLES]*, 41(9), pp. 954–954.

Maslach, C. and Jackson, S.E. (1981). The measurement of experienced burnout. *Journal of Occupational Behaviour*, 2, pp. 99–113.

McColl-Kennedy, J.R. and Anderson, R.D. (2002). Impact of leadership style and emotions on subordinate performance. *Leadership Quarterly*, 13, pp. 545–559.

McDowell, W.C. and Zhang, L. (2009). Mediating effects of potency of team cohesiveness and team innovation. *The Journal of Organizational Leadership and Business*, (Summer) pp. 1–11.

Miles, M. and Huberman, A (1984). *Qualitative data analysis. A sourcebook of new methods*. Beverley Hills, CA: Sage Publications.

Owen, C.A. (2001). The role of organisational context in mediating workplace learning and performance. *Computers in Human Behavior*, 17, pp. 597–614.

Patton, M.Q. (2002). *Qualitative research and evaluation methods*. London: Sage.

Reus, T.H. and Liu, Y. (2009). Rhyme and reason: Emotional capability and the performance of knowledge-intensive work groups. *Human Performance*, 17(2), pp. 245–266.

Roth, W.-M. (2007). Emotion at work: A contribution to third-generation cultural-historical activity theory. *Mind, Culture, and Activity*, 14(1–2), pp. 40–63.

Salas, E., Rosen, M. et al. (2007). Markers for enhancing team cognition in complex environments: The power of team performance. *Aviation, Space, and Environmental Medicine*, 78(5), pp. 77–85.

Stokes, C.K., Lyons, J.B. and Schneider, T.R. (2011). The impact of mood on interpersonal trust: Implications of multicultural teams. In *Trust in military teams*, edited by N.A. Stanton. Farnham: Ashgate Publishing. pp. 13–30.

Sturdy, A. (2003). Knowing the unknowable? A discussion of methodological and theoretical issues in emotion research and organizational studies. *Organization*, 10(1), pp. 81–105.

Tams, S. (2006). Constructing self-efficacy at work: A person-centered perspective. *Personnel Review*, 37(2), pp. 165–183.

Tasa, K. Taggar, S. and Seijts G.H. (2006). The development of collective efficacy in teams: A multilevel and longitudinal perspective. *Journal of Applied Psychology*, 92(1), pp. 17–27.

Tesch, R. (1990). The mechanics of interpretational qualitative analysis. In R. Tesch (ed.) *Qualitative research: Analysis types & software tools*. New York: Falmer Press, pp. 113–127.

The Australian Consice Oxford Dictionary (1987). Victoria, Australia: Oxford University Press.

Valsiner, J. and van der Veer, R. (2005). On the social nature of human cognition: An analysis of the shared intellectual roots of George Herbert Mead and Lev Vygotsky. In H. Daniels (ed.), *An introduction to Vygotsky*. Hove: Routledge, pp. 81–100.

Williams, S.A., Wissing, M.P. and Temane, Q.M. (2010). Self-efficacy, work, and psychological outcomes in a public service context. *Journal of Psychology in Africa*, 20(1), pp. 43–52.

Chapter 6

The Impact of Team Member Familiarity on Performance: Ad hoc and Pre-formed Emergency Service Teams

Dr Peter Hayes

Bushfire CRC and Kaplan Business School, Melbourne, Australia

Introduction

Teams have become the primary unit used to conduct emergency service activities. Fire, ambulance, police, search and rescue, and coastguard use teams extensively, ranging from frontline response to command, control and communication (C3) activities. Emergency service teams regularly operate in difficult environments. These teams are required to make decisions and act in dynamic environments that may be uncertain, time-pressured, involve high stakes and pose a threat to either the team or community members.

The challenges confronting emergency service teams may be complex and dynamic so that no single decision maker can develop an adequate understanding of all the issues. This has led to the development of distributed decision making where each team member takes responsibility for a component of the decision making (Brehmer 1991). Teams using distributed decision making are found across a range of organisations and in many other settings besides emergency services, including health, military and industry (Hollenbeck et al. 1995). Central to the composition of such teams is the extent to which members have worked with each other in the past and their knowledge of one another (i.e., member familiarity). This chapter explores the influence of member familiarity on teamwork processes and decision making.

Frequently emergency service organisations may need to deploy teams where the members haven't previously trained or worked together. Several important questions arise from this requirement. First, do teams consisting of personnel who have not worked together before (i.e., an unfamiliar or mixed familiarity team) perform as well as teams where all of the members have worked together previously (i.e., an intact or pre-formed team)? Second, if there are performance differences between unfamiliar and intact (familiar) teams: (a) what are these differences, (b) how large are they, and (c) are they of practical significance to emergency service agencies? Lastly, what types of intervention might an emergency service organisation use to help teams integrate members who haven't worked or trained together?

This chapter considers how the familiarity of members may affect team performance. The research informing this chapter comes from a variety of domains. Where possible, the research referred to has been undertaken with teams in naturalistic settings. Unfortunately, only limited research has been carried out specifically with emergency service teams. However, research from military, aviation, medicine and other organisational settings provides important insights on this issue. The first section of the chapter reviews relevant team and decision making research describing how teams develop, the nature of teamwork and taskwork, and team decision making, and identifies the specific teamwork mechanisms through which familiarity may influence team decision making and performance. This section also discusses research that specifically addresses the influence of team familiarity on team performance. In addition the implications of the research findings for team functioning and decision making are discussed.

The second section outlines the implications suggested by this research for emergency services managers. Two interventions are outlined that may help emergency service agencies support teams when members haven't worked together, namely brief résumés and brief question and answer sessions.

The third section describes some approaches that instructors can use to highlight the importance of familiarity for team performance. Two ideas are suggested, namely highlighting how member familiarity influences communication and coordination within teams, and the importance of a psychological safety (i.e., team members are able to speak up) in supporting information sharing and problem solving.

The chapter concludes with a summary of the key observations and offers suggestions for further reading on this topic.

An Outline of Relevant Team Research and Findings

Description of Teams

Teams involve interdependent members working in specific roles who interact adaptively to complete a common objective (Salas et al. 1992). Teams are dynamic and complex systems that evolve over time and are adaptive to situational demands (Kozlowski and Ilgen 2006). The key point of difference between teams and groups is the interdependency of team members, and it is interdependency that requires team members to work in a collaborative and coordinated manner to be effective.

Team Development and Lifespan

The systematic change in teams over time has been noted by researchers for over 50 years. Tuckman (1965) proposed a stage model suggesting teams progress through the phases of forming, storming, norming and performing. Since Tuckman's initial work a variety of alternative developmental models have been proposed for teams. The more recent models have moved away from a largely linear view of development indicated by stage models, suggesting that teams follow a variety of pathways. For example, Gersick (1988, 1989) suggested teams follow a pattern of 'punctuated equilibrium' moving through periods of inertia and behavioural changes that appear to be more influenced by members' awareness of time and deadlines than any particular period of time within a developmental stage. Kozlowski and Bell (2003) observed that teams are dynamic entities, and that team-level phenomena emerge from the individual level and develop in a complex way over time. This development is not only linear, but also includes episodic and cyclical aspects.

The Nature of Teamwork, Taskwork and Team Decision Making

Teamwork and Taskwork

For a team to operate effectively, members need not only sound technical knowledge, skills and attitudes, but also the capacity to cooperate and coordinate their actions with their colleagues. In essence there are two important types of behaviours required for a team to perform well, namely taskwork and teamwork (Morgan et al. 1986; McIntyre and Salas 1995). Taskwork involves the competencies – knowledge, skills, attitudes and other characteristics (KSAOs) – directly related to an individual's task performance, whereas teamwork involves the competencies (KSAOs) required for a member to operate effectively within a team (Salas et al. 2009). Teamwork facilitates taskwork and effective team performance requires the successful integration of taskwork and teamwork activities (Salas et al. 2005). See also Brooks (Chapter 9, this volume) for further discussion of emergency management competencies.

Drawing on aviation research focused on temporary (ad hoc) teams, Flin and Maran (2004) developed a teamwork model for acute medical teams. The authors identified four team skills central to performance: cooperation, coordination, leadership and communication (see Figure 6.1).

Team Decision Making

Team decision making has been defined as 'the process of reaching a decision undertaken by interdependent individuals to achieve a common goal' (Orasanu and Salas 1993, p. 328). Team decision making requires managing multiple information sources and varying task perspectives in order to reach a final decision (Ilgen et al. 1995). For example, teams contain members with unique histories, differing levels and types of expertise, and varying status levels (both formal and informal).

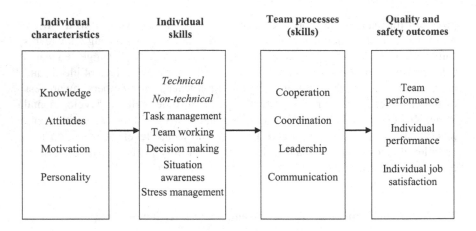

Individual characteristics	Individual skills	Team processes (skills)	Quality and safety outcomes
Knowledge	*Technical*	Cooperation	Team performance
Attitudes	*Non-technical*	Coordination	
	Task management		Individual performance
	Team working		
Motivation	Decision making	Leadership	
	Situation		Individual job satisfaction
Personality	awareness	Communication	
	Stress management		

**Figure 6.1   Factors affecting performance in teams (adapted from Flin
and Maran 2004, with permission)**

Note: Adapted by permission from BMJ Publishing Group Limited (Identifying and training
non-technical skills for teams in acute medicine; R. Flin and N. Maran 2004; *Quality and
Safety in Health Care*; 13:i80–i84.

Decision Making Research

The focus of classical (i.e., normative) decision research has been largely on
the individual decision maker. The early classical decision models assumed the
individual as a rational decision maker, seeking to maximise utility, and operating
in a reasonably static or constant decision environment. Organisational decision
models and naturalistic decision making models differ from such normative
models in that these approaches have been developed from observing decision
makers situated within teams or organisations. Most importantly, both naturalistic
decision making and organisational decision models focus on what decision
makers actually do in teams and organisations (Lipshitz, Klein and Carroll 2006),
and thus take account of the often dynamic environment within which decision
makers are required to operate.

Organisational Decision Models

Herbert Simon recognised that people working in organisations frequently
simplified their decision making by limiting their consideration of an issue. By
using rules of thumb or heuristics, decision makers are able to consider an issue
within a 'bounded rationality', making decisions that satisfy certain criteria (i.e.,
satisficing decisions) rather than necessarily offering the greatest overall return
(Simon 1955, 1979). Organisational decision models have particular relevance
for teams who tend to be guided by organisationally determined processes and

decision rules, e.g., standard operating procedures (SOPs). These teams often operate in time-constrained situations, sometimes necessitating the selection of the first satisfactory option rather than risk waiting for the optimal option which may or may not occur.

Naturalistic Decision Making

The naturalistic decision making framework was developed from observing expert decision makers (i.e., fireground commanders) in uncertain, dynamic, time-pressured and often high-stakes circumstances. Unlike classical decision theories, which tend to concentrate on how to choose between different options, naturalistic decision making focuses on the assessment and appropriate categorisation of the situation (Klein 1997). Whilst a novice decision maker tends to rely on his/her reasoning skills to make decisions, experts' decisions are based on their situation assessment abilities (Burke et al. 2008). A further feature of naturalistic decision making is its treatment of decision making cycles, suggesting that these are often interwoven with action. Instead of decision makers gathering and analysing all of the relevant data and then making a decision, people in complex situations often 'think a little, act a little, and then evaluate the outcomes and think and act some more' (Orasanu and Connolly 1993, p. 19). See also Bremner, Bearman and Lawson, Chapter 8, and Brooks, Chapter 9, this volume for further discussion of naturalistic decision making in emergency management.

Distributed Decision Making in Emergency Service Teams

The dynamic environment that emergency service teams may be required to operate within was noted in the introduction. Brehmer (1992) described four features of dynamic decision making environments that make them particularly challenging:

1. a series of decisions is required to reach the goal
2. the decisions required are not independent; earlier decisions tend to constrain later decisions
3. the state of the environment continues to change, both autonomously and as a consequence of the decision maker's actions
4. decisions need to be made in real time.

Teams operating in dynamic decision making environments are typically under considerable time pressure and may need to make decisions under conditions of the uncertainty associated with:

1. when a team may receive information (Clancy et al. 2003)
2. information accuracy (Orasansu and Connelly 1993; Schmitt and Klein 1996)

3. information completeness (Orasansu and Connelly 1993; Schmitt and Klein 1996)
4. information complexity (Schmitt and Klein 1996; Lipshitz and Strauss 1997).

Clancy et al. (2003) suggested that in a dynamic environment, the timeliness of decision making can be as important as the decision content.

The interdependency of team members is an important feature that differentiates a team from a group. Groups tend to make decisions on a consensus basis, with each member using the same information. In contrast, the interdependent nature of teams can lead to distributed decision making, where each team member takes responsibility for a component of the decision making (Brehmer 1991). This requires team members to collaborate in developing and maintaining a shared situation awareness of developing events. Distributed decision making is commonly used in dynamic decision making environments such as firefighting, medical emergencies and military combat operations (Clancy et al. 2003). We can also conceptualise a team as a cognitive unit. Some researchers use the term *team cognition* to capture the various collective processes (e.g., communication, decision making, memory, perception and situation awareness) required by a team. The distributed nature of how information is analysed and decisions are made in teams increases the requirements to coordinate how members share information and develop an appropriate level of team situation awareness.

How Does Team Familiarity Affect Team Performance?

Familiarity is generally positively related to various aspects of effective team performance in a variety of settings (e.g. Reagans, Argote and Brooks 2005; Espevik et al. 2006; Huckman, Staats and Upton 2009).

Reagans, Argote and Brooks (2005) suggested that the factors which are used to explain the influence of familiarity on team performance can be divided into two complementary groups:

1. mechanisms that support team coordination
2. mechanisms that influence the effective interaction and relationships of team members.

Familiarity positively influences a team's ability to act in a coordinated fashion (Moreland, Argote and Krishnan 1998). Through shared experience Huckman, Staats and Upton (2009) suggested that individual members develop team-oriented customs and norms and that these in turn aid coordinated performance (Mailath and Postlewaite 1990; Chillemi and Gui 1997). Additional support for the role of familiarity in helping coordinate team performance comes from Weick and Roberts (1993) who proposed that familiarity is important in enabling 'heedful interrelating', significant in explaining the almost faultless performance of an aircraft carrier flight deck. Team members interrelate heedfully when they consider the big picture and their contribution to the collective goals in a careful, critical and purposeful manner. It takes time to develop the shared understanding, openness and disclosure required to heedfully interrelate, and as Weick and Roberts point out, these are indicators of a mature and well-developed team.

Perhaps the most parsimonious explanation of how familiarity may influence team performance is the McLennan et al. (2006) observation that familiar teams tend to use their cognitive resources more effectively. This observation suggests that familiarity may influence team performance through interpersonal relationships and the knowledge of one another that tends to develop with familiarity, and that these relationships and team knowledge enable superior team coordination.

However, there is some evidence that teams that continue to work together for extended periods may show curvilinear patterns of performance with decrements in performance in the longer term (e.g., Katz 1982; Berman, Down and Hill 2002). Katz observed that team performance began to diminish after the project teams had been together for five years. Katz and Allen (1982) described this as the 'not-invented-here' (NIH) syndrome, where long-serving teams believe they have a monopoly of knowledge and thus reject new ideas from outsiders to the likely detriment of team performance.

Research conducted with US Army helicopter crews by Leedom and Simon (1995) found mixed results for the performance of highly familiar (battle rostered) versus mixed familiarity crews. Although the battle rostered crews tended to achieve more tactical simulator missile kills (57/65) than the mixed familiarity crews (46/66), the battle rostered crews tended to have a greater degree of overconfidence than the mixed familiarity crews.

Other researchers have failed to find evidence for the curvilinear relationship between team tenure and team performance. The Keller (1986) study of 32 research and development project teams did not find a curvilinear relationship between team tenure and performance. The best predictors of team performance in the Keller study were team cohesiveness and an innovative orientation as a team. Similar to the Keller study, the Allen et al. (1988) study of 181 research and development teams from nine organisations found no evidence of a curvilinear relationship between team tenure and performance. Although Allen et al. found there was some variability in the performance of the long-tenured teams, there was no general decay in the pattern of performance over time. The authors investigated the relationship between the leadership attributes of the research and development teams' functional and project managers, and team performance. Allen et al. found that the leadership attributes most useful for supporting good performance were somewhat different between the newly formed and long-tenure teams (see also Owen, Chapter 7, this volume).

Perhaps the most plausible interpretation of the curvilinear relationship found between team tenure and team performance is that it is not tenure per se that influences team dysfunction. How well a team functions depends on a combination of team norms, leadership, context and the characteristics of team members.

However, an implication of the Allen et al. (1988) study is that there may be periods during a team's existence when a team is more susceptible to dysfunction. The team tenure research also reminds us that team member familiarity in no way guarantees good performance and that the development of poor team norms and a lack of leadership can be problematic for team performance.

Key Team and Teamwork Processes Influenced by Team Familiarity

Earlier the crucial contribution that teamwork makes to team performance was described. This section describes the three important teamwork processes (i.e., communication, coordination and leadership) and the emergent team process of trust. These processes are influenced by team familiarity and help explain why familiarity is likely to influence effective team performance. These four processes are interrelated. For example, effective leadership is likely to positively influence team coordination, good communication will positively influence coordination, effective leadership may positively influence communication, and trust positively influences team communication.

Team Communication

Communication enables team members to exchange information, coordinate activities, monitor performance, provide feedback, create plans and strategy, and develop a shared understanding of the incident (Rasker, Post and Schraagen 2000). Communication plays a major role in team coordination and importantly, in helping teams develop situation awareness (Palmer 1990). As familiarity increases in teams, communication between team members tends to become more coordinated (Reid and Reed 2000).

The distributed dynamic decision making (D3M) that occurs within emergency service teams heightens the need for effective communication. For example, although simple incidents may be considered relatively routine, as incidents (e.g., hazardous chemical spill, cyclone, or riot) increase in scale and complexity they quickly become non-routine and intra-team communication becomes central to team performance in these conditions (Orasanu and Salas 1993).

MacMillan et al. (2002) made an important observation regarding team coordination and communication, noting that where a team requires communication to coordinate its activities, the need to communicate can have a negative impact on team performance, especially in high workload conditions. MacMillan et al. also noted that 'communication can be good or bad for team performance, depending on when it occurs and what else is going on at the time' (p. 293).

Edmondson (1999) highlighted the importance of psychological safety, a team climate where members will speak up and offer constructive comment to their team mates. Psychological safety develops in teams with shared experience and involves the development of trust and mutual respect within the team (Edmondson 1999). Leadership in the form of leader inclusiveness has also been shown to influence the development of psychological safety (Nembhard and Edmondson 2006). Leader inclusive behaviours invite and show appreciation for team member contributions. When psychological safety has developed in a team, members will be confident that they will not be undermined, penalised, or embarrassed by their colleagues for speaking up, asking questions, proposing new ideas, or reporting an error (Edmonson 2004; Edmonson and Roloff 2009). Moreover, psychological safety has been shown to support team learning, providing the ongoing opportunity to improve team performance (Edmonson, Bohmer and Pisano 2001).

Research conducted by Kanki and Foushee (1989) found that familiar flight crews demonstrated a more open approach to communication, exchanging more information than crews who had not flown together before. This research suggests that familiar team members are more likely to provide constructive comment to one another and listen more carefully to comments and suggestions from fellow team members.

Coordination

Coordination is seen as a cornerstone of effective team performance (Zalesny, Salas and Prince 1995; Klein et al. 2005). In discussing why teams often fail to perform against expectation, Steiner (1972) developed the term *process loss* to describe performance costs associated with poor team coordination. Team coordination involves 'team members executing their activities in a timely and integrated manner' (Schraagen and van de Ven 2011, p. 178). Familiar teams tend to have developed a greater repertoire of mechanisms (e.g., shared knowledge and customs) to coordinate their activities than unfamiliar teams.

Klein et al. (2005) suggested that there are three primary requirements for team coordination: interpredictability, common ground and directability. For a team to coordinate their actions it is important that members can predict the actions of other members with some degree of accuracy (i.e., interpredictability). Team members also need to ensure that their own actions are sufficiently predictable to support effective coordination. Klein et al. noted that predictability can enable team members to accurately estimate important features of the situation. For example, how much time is required to complete key tasks, the level of skill required, and the degree of complexity and difficulty of tasks. Common operating frameworks such as incident control systems and SOPs support interpredictability by creating expectations of how team members will behave. Douglas (Chapter 5, this volume) highlights the importance of member predictability, particularly when teams are working in emotionally challenging situations.

Clark and Brennan (1991) observed that common ground is a fundamental requirement for coordinated action. Common ground refers to 'the pertinent mutual knowledge, mutual beliefs, and mutual assumptions that support interdependent actions in joint activity' (Klein et al. 2005, p. 146). Common ground can support efficient communications within a team. For example, it can enable team members to abbreviate intra-team communications yet still be confident that potentially ambiguous messages will be correctly understood.

The third requirement for team coordination is the ability for team members to be able to redirect each other's actions (Christoffersen and Woods 2002). Directability is the ability to modify the actions of other team members as conditions and priorities change. This may occur when a team member noticing that another team member has run into difficulties, alters their own activities to compensate. It may also occur when a team member signals to the wider team that they are either ahead or behind schedule on a key task so that their colleagues can make suitable adjustments. The capacity to redirect enables teams to perform in a more resilient way, able to cope with changing conditions.

The Wittenbaum, Vaughan and Stasser (1998) circumplex model suggested that teams can coordinate their activities along two dimensions: process and temporal. In terms of team processes, activities can be coordinated:

1. explicitly through activity programming (e.g., plans and procedures), and via communication (e.g., oral, written, team meetings)
2. implicitly through shared knowledge of the situation, tasks at hand and each other (MacMillan, Entin and Serfaty 2004).

A particular advantage of implicit coordination is its lower overhead, involving less member time and cognitive capacity, and thus assisting team performance (MacMillan, Entin and Serfaty et al. 2004).

A second way that team coordination can be considered is on a temporal basis (Fiore et al. 2003; Wittenbaum, Vaughan and Stasser 1998). Given the complex and dynamic environments that teams operate within, information overload can be a particular problem for team coordination (Fiore et al. 2003). Technology adds to this challenge by often providing large amounts of information (Klein 2009; Edmunds and Morris 2000). Teams can coordinate and manage information during three different periods of time:

1. prior to the team activity occurring (pre-process)
2. during the team activities (in-process)
3. after the team activities have occurred (post-process).

The Wittenbaum, Vaughan and Stasser (1998) and Klein et al. (2005) frameworks provided somewhat complementary accounts of the factors influencing team coordination. The Klein et al. account suggested that team members who have worked together tend to develop an informal understanding of one another's skills, habits and way of working. In essence, common ground is a form of intra-team situation awareness that overlaps with the Wittenbaum, Vaughan and Stasser concepts of pre-process and implicit coordination. To enable crews or incident management teams to function immediately, organisations provide comprehensive training, protocols, checklists and a variety of SOPs. These organisational preparations are examples of pre-process coordination that explicitly help support greater team member interpredictability and common ground. The concept of directability (Klein et al. 2005) is an example of an in-process coordination mechanism (Wittenbaum, Vaughan and Stasser 1998) that may also be thought of as a form of leadership. In the following section, the role of leadership in team performance is discussed. The importance of shared leadership, which may also support the coordinating mechanism of directability, is described.

Although coordinated performance in emergency response teams tends to be chaotic early on, it tends to become self-organised over time (Guastello 2010). Recent simulation-based research conducted by Gorman and Cooke (2011) with three-person crews flying unmanned aerial vehicles indicated that team cognition evolves in an exponential way as crew members become increasingly familiar. Gorman and Cooke (2011) observed that as members become familiar there is a shift in the basis of team cognition from interaction to shared knowledge.

Leadership

Two types of leadership appear to be most relevant to teams: vertical leadership and shared leadership (Pearce 2004). Vertical leadership is a traditional model of leadership which suggests the senior officer or the manager is in charge whilst the other team members are simply followers. Shared (or distributed) leadership occurs when team members assist in providing leadership, and this assistance varies over time depending on the particular issues challenging the team and the specific competencies of team members. Shared leadership takes time to develop in teams and generally requires some degree of familiarity. Pearce and Conger (2003) defined shared leadership as 'a dynamic, interactive influence process among individuals in groups for which the objective is to lead one another to the achievement of group or organizational goals or both' (p. 1).

Pearce (2004) and colleagues (Conger and Pearce 2003; Cox, Pearce and Perry 2003) argued that both vertical and shared leadership are important for effective team performance. Cox, Pearce and Perry (2003) suggested that shared leadership supplements vertical leadership and that vertical leadership is an antecedent to shared leadership in teams. In a study of 71 change management teams, Pearce and Sims (2002) noted that although both vertical and shared leadership were important predictors of team performance, shared leadership was found to be the more useful predictor.

The type of leadership required changes as teams mature and with the level of expertise within a team (Kozlowski et al. 2009). This observation suggests that the leadership style in unfamiliar (ad hoc) teams may differ to that found in familiar (intact or pre-formed) teams and may also vary depending on the level of expertise found within the team. The development of shared leadership appears to be related to several aspects of team familiarity. Shared leadership develops over time as team members interact, influence one another, and negotiate their roles within the team (Carson, Tesluk and Marrarone 2007). It takes time for team members to develop an understanding of their respective competencies and thus familiar teams are more likely to develop the capacity and willingness required to engage shared leadership than unfamiliar teams (Perry, Pearce and Sims 1999; Avolio et al. 1996).

Trust

A further way that familiarity may influence team relationships and performance is through trust. DeJong and Elfring (2010) defined trust as 'a psychological state involving confident, positive expectations about the actions of another' (p. 536). Shared experience can lead to trust, an important factor influencing the sharing of information within a team (McEvily, Perrone and Zaheer 2003). Liang, Moreland and Argote (1995) found that teams who have trained together were more likely to trust one another's expertise. Trust has been found to be critical in a range of emergency situations, especially where time is critical and the response requires good coordination (Omodei, Wearing and McLennan 2000; Omodei and McLennan 2000).

Trust has been shown to favourably influence a number of important team processes (Adams and Webb 2000). Trust within a team can reduce the need for monitoring of other members' performance (McAllister 1995; Currall and Judge 1995) and lead to improved cooperation (Meyerson, Weick and Kramer 1996; Rousseau et al. 1998). Nahapiet and Ghoshal (1998) emphasised the relationship between trust and cooperation, noting that 'trust lubricates cooperation and cooperation itself breeds trust' (p. 255). Higher levels of trust facilitate within-team communication (O'Reilly 1978; Currall and Judge 1995), reduce conflict between members (McEvily, Perrone and Zaheer 1998) and improve team processes and performance (Dirks 1999; Kirkman et al. 2006). However, low levels of trust can be problematic in teams because this can add costs that reduce team effectiveness (Wilson, Straus and McEvily 2006).

Adams and Webb's (2003) research suggested that there are two types of trust at play during the formation and operation of a team. The first type of trust is person-based trust, which develops over time and is based on prolonged interactions, enabling building the knowledge and experience of others necessary for trust (Rempel, Holmes and Zanna 1985; Lewicki and Bunker 1996). The second form of trust is category-based trust, which develops on the basis of a person's membership of a group (or category) that is positively linked with trust (Kramer 1999). Category-based trust can occur even when circumstances do not allow for the development of person-based trust.

Swift trust is a particular form of category-based trust that enables a range of temporary teams to operate successfully. Meyerson, Weick and Kramer (1996) observed that diagnostic teams, film crews, aircraft cockpit crews, paramedics and firefighting teams all operate effectively even though their members may have not worked together previously. For example, a new team member's previous experience working in a known high-performing team would help provide confidence for colleagues that this new member is competent and can be trusted.

In considering the role of trust in team functioning it is important to take account of the appropriate level of trust for a given situation: more trust isn't necessarily better (McEvily, Perrone and Zaheer 2003). An important judgement that team members need to make is when do they trust the recommendations of colleagues and personnel outside the team and when do they ask for more information from them? Trust is also context specific, so a colleague's judgement may be readily accepted for a particular situation, but might be questioned in different circumstances.

McEvily et al. (2003) offered an alternative view of trust, proposing that it is an organising framework or heuristic that helps select and guide behaviours. A heuristic trust offers certain efficiencies, for example allowing team members to economise on information processing and monitoring, thereby conserving cognitive resources in information-rich but time-poor environments such as emergency incidents. However, where team members rely too heavily on trust, this may limit their scanning of wider networks and alternative information sources that might lead to improved decision making (McEvily et al. 2003).

The Influence of Familiarity on Team Performance

In this section four studies that consider how familiarity influences team performance are discussed. The studies come from the aviation, medicine, military and emergency management sectors. A common feature of the teams in these studies was the requirement to work in a highly coordinated manner in order to be successful.

The National Transportation Safety Board (NTSB) (1994) used archival data to investigate whether team familiarity was a factor in flight crew-involved major crashes of US commercial flights between 1978 and 1990. The NTSB found 44 per cent of accidents occurred on the first leg of newly paired flight crews (i.e., where the pilot and co-pilot have not flown together before). The incidence of crews involved in accidents during their first flight together was considerably greater than the estimated proportion of new flight crew pairings (ranging from 2.8 per cent to 10.5 per cent). The NTSB found that 73 per cent of accidents occurred during the first day of a flight crew pairing. Similarly, the incidence of crews involved in accidents during their first day together was considerably greater than the estimated proportion of new flight crew pairings (ranging from 6.8 per cent to 30.3 per cent). The NTSB study suggested that low team familiarity may lead to poorer decision making and pose greater operational risks.

Reagans, Argote and Brooks (2005) investigated surgical teams undertaking joint replacements over a five-year period. These teams were rostered to work together so that they may be working with either familiar or unfamiliar peers. The researchers assessed the performance of the orthopaedic teams for 1,151 hip and knee replacements. Familiar teams carried out the joint replacement operations more quickly than the unfamiliar teams. For example, a surgical team that had undertaken 10 knee replacements together would take 5 per cent less time (approximately 10 minutes) to perform this operation than a surgical team that had not worked together before.

Espevik et al. (2006) compared the performance of six-person submarine attack teams in a tactical submarine simulator. In one simulation, the participants worked in intact teams (i.e., having worked together as crew on a submarine for the three months prior to the experiment), and in the second simulation, the second-in-charge for each team was swapped with a person unfamiliar with the rest of the team. The intact teams scored 33 per cent more hits ($M = 4$) than the mixed familiarity teams ($M = 3$). The authors also noted that there was a trend for the familiar teams to discover, identify, attack and hit targets at greater distance than the mixed familiarity teams.

Hayes and Omodei (2012) used simulation-based research to compare the performance of small (four-person) unfamiliar (ad hoc) and pre-formed (familiar) wildfire incident management teams. The authors found that, compared to ad hoc (unfamiliar and mixed familiarity) teams, the pre-formed teams showed superior performance across a range of measures including:

1. the number of fireground events attended to
2. the quality of responses to fireground events
3. the quality of handover briefings and documents
4. timeliness of decision making
5. the level of team situation awareness
6. team and teamwork processes (e.g., intra-team communication, coordination, leadership and trust).

The pattern of results indicated a positive exponential relationship between member familiarity and team performance. The authors noted that the familiar (pre-formed) teams performed better than the ad hoc (unfamiliar and mixed familiarity) teams both in terms of the number of simulated fireground events that they attended to (efficiency) and the quality of these responses (thoroughness). Importantly, the familiar teams managed to perform better on both counts than the unfamiliar teams, suggesting that their greater level of coordination enabled them to deal more effectively with the fireground events. This observation is consistent with the McLennan et al. (2006) observation of greater cognitive efficiency meaning that the familiar teams' limited resources went further than the unfamiliar teams'.

The Hayes and Omodei (2012) study used two intra-team communication measures: listening to other team members and providing constructive comment to other team members. The higher levels of constructive comment and listening to team members in the familiar teams suggest that psychological safety was an important aspect of team climate that positively influenced team performance. Recent research by Lewis, Hall and Black (2011) highlights the importance of psychological safety in teams, observing that even experienced firefighters may face social pressures that mean they remain silent in some situations (see also Owen, Chapter 7, this volume).

An interesting question prompted by the curvilinear results from the Hayes and Omodei (2012) study is whether there is more at play than simply member familiarity shaping team performance? In other words, does pre-training teams (i.e., pre-formed) provide an incremental benefit over and above familiar team members working together? The pre-training of teams is designed to enable teams to rapidly commence management of an incident in a seamless manner. The term *familiar* suggests that team members have developed some degree of rapport. Depending on the nature and duration of their previous shared experiences, familiar team members may also have developed some understanding of each other's backgrounds and relevant work experience. However, familiarity by itself does not necessarily ensure team members have sufficient understanding and knowledge of one another in specific roles that will support high levels of team coordination. The pre-training of pre-formed teams can help ensure clarity of member roles and responsibilities, develop the capacity to use implicit coordination and ensure they are able to allocate suitable work to members. In sum, pre-training supports greater member interpredictability than simple familiarity and thus may further aid team coordination and performance.

It was noted earlier that team performance is influenced by mechanisms that support:

1. team member relationships and interaction
2. team coordination.

Team member familiarity appears most likely to support (1) and some aspects of (2). Effective pre-training should not only support member familiarity, but also assist in some team coordination issues not necessarily addressed by familiarity (e.g., member roles and responsibilities). Particularly in time-critical settings, the greater clarity of team member roles and responsibilities of pre-formed teams may provide a valuable performance advantage over teams comprising members who are simply familiar with one another.

A second interesting familiarity-related issue for the performance of teams is whether there is enhanced learning from members rotating or working in other teams. An important finding from Hayes and Omodei (2011) was the requirement for adaptive and flexible incident management personnel who can improvise. Pre-training and pre-forming teams tends to create greater predictability for team members, which generally should support superior team coordination and performance. However, this predictable environment does not necessarily create the opportunities to regularly work with new colleagues and the variation important in the development of adaptive and flexible personnel.

The Gorman and Cooke (2011) simulation-based research with teams operating unmanned aerial vehicles has interesting implications for how familiarity may influence team cognition. The authors found that mixing members from different teams resulted in significant knowledge and process gain. In other words, there were team learning benefits from mixing teams. The authors also observed that as team members became more familiar, cognition in teams shifted from an interactive basis (i.e., team interaction is team cognition) to a greater emphasis on shared knowledge (i.e., team mind). Moreover, this transition in team cognition from interaction-based to shared knowledge occurred in a non-linear manner that may be best described by nonlinear dynamics (Cooke and Gorman 2009; Gorman, Amazeen and Cooke 2010). Nonlinear dynamical systems is a general systems theory used to describe and predict change processes (Guastello 2010). Nonlinear dynamical systems may appear unpredictable or random, but are essentially deterministic. A team's start point will influence subsequent shifts in team cognition and coordination. For example, in wildfire incident management teams, this may be the combination of the level of member familiarity and the state of the incident (e.g., uncontained small fire vs. contained large fire).

In terms of the Hayes and Omodei (2012) study, the high workload requirements may have disadvantaged the less familiar (ad hoc) teams reliant on interactive cognitive processes and favoured the familiar (pre-formed) teams using more shared cognitive processes. Moreover, the Gorman and Cook (2011) research suggests that changing team membership may help team learning for future settings – despite the consequential lower member familiarity – but undermine team performance in the current setting, especially in high workload D3M settings.

Figure 6.2 integrates the findings from Hayes (2012) with the Gorman, Amazeen and Cooke (2010), Gorman and Cooke (2011) and the Reagans, Argote and Brooks (2005) perspectives of how team familiarity influences team performance. In a manner consistent with the Hayes (2012) findings, Gorman, Amazeen and Cooke suggested that team cognition evolves in a positive exponential way as members become increasingly familiar. This evolving team cognition, which is based on increasing shared knowledge, may help support greater team efficiency. The nonlinear dynamical systems characterisation of team cognition suggests that the shift from a dominance of interaction to shared knowledge may occur at various points in time as well as being dependent on the system starting point. The upward concave curve in Figure 6.2 approximates the Hayes (2012) results and is suggestive of the general relationship proposed by Gorman, Amazeen and Cooke. A pre-trained team is likely to commence operating at a higher level of efficiency than a merely familiar team. Figure 6.2 suggests that the Reagans, Argote and Brooks mechanisms of member interaction and relationships, and team coordination also tend to evolve in a team. In the early stages of a team whose members are new to one another, member interaction and relationship development will tend to be prevalent. As the team trains or works together, there will be an increasing focus on coordinating team and member activities. This evolving pattern of activities can be viewed as behavioural indicators of the changes in team cognition.

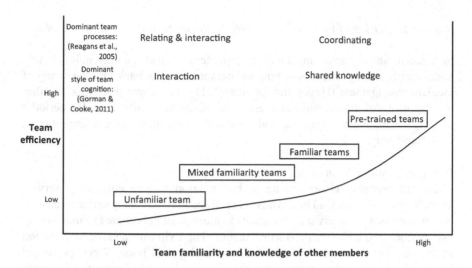

Figure 6.2   Conceptual view of the Hayes (2012) findings integrated with the mechanisms influencing team performance

An implication of Figure 6.2 is that to achieve maximum levels of efficiency requires teams to be pre-trained. In the case of incident management teams, although these teams develop high levels of familiarity by working together, the infrequent deployment of these teams and membership changes suggests that pre-training in regular exercises may be the best way for emergency service agencies to ensure very high levels of team efficiency on an ongoing basis.

Implications for Practitioners and Instructors

The evidence provided in this chapter suggests that teams which regularly work together tend to be more effective than ad hoc teams. The apparent greater cognitive efficiency of familiar teams suggests that these teams will be particularly valuable for managing difficult tasks or incidents.

Value of Mixing Personnel from Different Teams to Enhance Training

The Gorman and Cooke (2011) research suggested that there may be learning benefits from conducting team-based training and exercises with personnel from different teams. Requiring members to work in an unfamiliar team can create an environment which may require personnel to interact more and to adapt to the likely variation in the way that their new colleagues conduct themselves.

Regular Exercising of Personnel who may be Required to Work with Each Other

In a recent study identifying incident management team competencies, it was noted that it usually takes some time for personnel to get back into the swing of incident management (Hayes and Omodei 2011). For teams that work together infrequently this observation suggests that agencies should consider periodic training to maintain teamwork and taskwork competencies, and importantly, team familiarity.

Increasing familiarity in ad-hoc teams
There are probably always going to be situations where emergency service agencies need to deploy ad hoc teams. Therefore, an important question to answer is how does an agency assist ad hoc teams to more quickly become familiar so they may perform more like the pre-formed teams? The swift trust and social cognition literatures provides some insights for considering this issue. When personnel haven't worked together, there is no experience to base judgements of another team member's competency, reliability and integrity (i.e., trustworthiness). In time-pressured situations, team members may rely on surface-level information in the form of category-based assessments of a team member who they have not worked with before. Meyerson, Weick and Kramer (1996) described this as importing expectations from other settings. Team members are particularly likely to develop category-based impressions of unknown personnel in time-constrained situations when little other information is available (Fiske and Neuberg 1990). Clearly this may be the case during the initial phases of a team's deployment where personnel haven't worked together previously.

There are two types of intervention that may assist members of ad hoc teams to work more effectively together – brief résumés and short question and answer sessions. The aim of these interventions is improve the ability of team members to coordinate their actions and this is achieved through developing team member knowledge of each other and fostering effective working relationships. These interventions may work in part by providing additional and categorical information that supports the development of swift trust.

Brief résumés provide the opportunity for managers and team members to rapidly appraise the likely capability of personnel who they may not have worked with before. Developing good teams is not only about placing good performers in the key roles, but also ensuring that personnel don't end up in the wrong role and thus undermine team performance. In a study of team member turnover, Levine and colleagues (2005) found that providing team members with information about newcomers task-relevant skills helped support higher levels of transactive memory (i.e., knowing who knows what in a team) and team performance, whether or not subsequent team turnover occurred. The Levine et al. study suggests that brief résumés may help teams more quickly and successfully integrate newcomers. A brief résumé may provide information about common connections or experiences and thus support the development of swift trust (Wildman et al. 2012).

A simple method that some experienced team leaders use to help assess personnel that they haven't worked with before is to ask a few simple questions about the person's experiences of working in teams. This may be done in a reasonably informal way, as a brief chat, and enable sectional leaders to gather information about the capability of personnel by asking about the last two or three incidents that they have worked on. Questions typically probe the person's role and responsibilities, the nature of the incident, the size of the team, their team or section leader, and how comfortable they felt in undertaking their duties. The more experienced team leaders use their knowledge of who the person has worked with to help assess the unfamiliar team member. For example, if an unfamiliar team member has successfully worked under a section leader who they know is a highly effective albeit demanding manager, then the likelihood is that the new person is a sound team member. Some simple probing can provide a reasonable sketch of the likely capabilities of a new team member and any local knowledge they may have.

The brief question and answer (Q&A) session fulfils two main functions. First, it provides further information about the likely capability of the unfamiliar team member and thus should help ensure the person is allocated to a suitable role. The Q&A may provide a more nuanced view of the likely competence of the person and thus assists in development of an appropriate level of trust. Similar to the brief résumé, the Q&A may provide information about common connections or experiences and enable the development of swift trust (Wildman et al. 2012). If there are doubts about the capability of the unfamiliar team member, then appropriate oversight of their work can be implemented. This approach may work by using trust as a heuristic, thus reducing the cognitive workload for team members by identifying where monitoring may or may not be required.

A second function a Q&A session may play is developing rapport between the unfamiliar team member and their new colleagues. Researchers highlight the central role that high-quality social relationships play within teams in supporting knowledge integration. A simple conversation is likely to help newcomers feel a little more at ease (i.e., psychologically safer) in the new team environment and thus more willing to make helpful suggestions that may assist the team to function effectively.

Strategies for Instructors

Highlighting the Relationship between Team Familiarity, Team Communication and Team Coordination

An important feature of team performance identified in this chapter is the close relationship between team communication and coordination. The influence of team familiarity is likely to be particularly valuable for team coordination and communication in time-constrained, high-workload settings. To highlight this communication-coordination relationship, instructors can use a novel short role-play exercise where participants are required to manage an incident with a mixture of familiar and unfamiliar team members. The idea here is to highlight the increased coordination and time costs associated with working with personnel who participants are unfamiliar with and identify some strategies to help better manage this. A short after-action review (AAR) following the exercise should help highlight some of the issues encountered by the teams and the opportunity to discuss strategies that can be used in these types of situations. For example, what can be done prior to deployment (pre-process coordination – e.g., exercising with new team members, brief résumés), during the incident (in-process coordination – e.g., sharing workload, brief question and answer), and consolidating the learning from the event (post-process coordination – e.g., AARs).

Supporting Psychological Safety in Teams

In order to encourage team members to speak up it is important that personnel understand the importance of psychological safety and some of the ways that they can help create and maintain this in team settings. By demonstrating behaviours that support psychological safety, an instructor can help create a safe training environment where participants can discuss and practise these same supportive behaviours. For example, the instructor may break the ice by highlighting the importance of recognising the fallibility of human decision making and the possible error traps that can be experienced. The instructor may then reveal to the group at a time when they have made an error or mistake on (e.g., a time that they became lost). By demonstrating openness, candour and perhaps some light humour, the instructor is modelling the types of behaviour that help develop rapport and trust in teams, and a climate that is psychologically safe.

The next step in this exercise would be to organise participants into groups where they take turns at talking about a time when they have made an error or mistake and discussing what they learned from that experience. The instructor could ask the groups to suggest ideas as to how they would develop rapport in a team where they didn't know anyone, and how they might help a new team member feel psychologically safe in a team. The last part of the group discussion could explore what prevents team members from speaking up and the importance for team leadership of adopting inclusive behaviours.

Conclusion

Emergency service teams are regularly required to work in difficult, complex, uncertain and dynamic environments. To perform well these teams need to coordinate their activities through teamwork so that effective and timely decisions can be made. The evidence presented in this chapter indicates that team communication, coordination, leadership and trust are central in enabling emergency service teams to effectively respond to key events, develop and implement appropriate plans, and make good quality decisions to resolve the incident. There will always be circumstances where team members haven't trained or worked together. By understanding how familiarity is likely to influence team performance, we can prepare teams and team members to better manage any unfamiliarity in these situations.

Suggestions for Further Reading

Flin, R., O'Connor, P., and Crichton, M. (2008). *Safety at the sharp end: A guide to non-technical skills.* Aldershot: Ashgate.

Guzzo, R.A., Salas, E. and Associates (eds) (1995). *Team effectiveness and decision making in organizations.* San Francisco: Jossey-Bass.

Hollnagel, E. (2009). *The ETTO principle–efficiency-thoroughness trade-off: Why things that go right sometimes go wrong.* Farnham: Ashgate.

Klein, G. (1999). *Sources of power: How people make decisions.* Cambridge, MA: MIT Press.

Swezey, R. and Salas, E. (1992). *Teams: Their training and performance.* New York: Ablex.

West, M.A. (2012). *Effective teamwork: Practical lessons from organizational research* (3rd edition). Chichester: BPS Blackwell.

Acknowledgements

The research was supported funding through the Bushfire Cooperative Research Centre. However, the views expressed are those of the author and do not necessarily reflect the views of the Board of the funding agency.

References

Adams, B.D., and Webb, R.D.G. (2002). *Trust in small military teams.* Paper presented at the 7th International Command and Control Research and Technology Symposium, Quebec City.

Adams, B.D., and Webb, R.D.G. (2003). *Trust development in small teams.* DRDC No. CR-2003-016 Toronto, Department of National Defence.

Allen, T., Katz, R., et al. (1988). Project team aging and performance: The roles of project and functional managers. *R&D Management*, 18, pp. 295–308.

Avolio, B.J., Jung, D., et al. (1996). Building highly developed teams: Focusing on shared leadership processes, efficacy, trust, and performance. In M. Beyerlein, D. Johnson, and S. Beyerlein (eds), *Advances in interdisciplinary studies of work teams: Team leadership* (Vol. 3, pp. 173–209). Greenwich: JAI.

Berman, S.L., Down, J., and Hill, C.W.L. (2002). Tacit knowledge as a source of competitive advantage in the National Basketball Association. *The Academy of Management Journal*, 45, pp. 13–31.

Brehmer, B. (1991). Distributed decision making: Some notes on the literature. In J. Rasmussen, B. Brehmer and J. Leplat (eds), *Distributed decision making: Cognitive models for cooperative work* (pp. 3–14). Chichester: Wiley.

Brehmer, B. (1992). Dynamic decision making: Human control of complex systems. *Acta Psychologica*, 81, pp. 211–241.

Burke, C.S., Priest, H.A., et al. (2008). Stress and teams: How stress affects decision making at the team level. In P.A. Hancock and J.L. Szalma (eds), *Performance Under Stress* (pp. 181–208). Aldershot: Ashgate.

Carson, J.B., Tesluk, P.E., and Marrone, J.A. (2007). Shared leadership in teams: An investigation of the antecedent conditions and performance. *Academy of Management Journal*, 50, pp. 1217–1234.

Chillemi, O., and Gui, B. (1997). Team human capital and worker mobility. *Journal of Labor Economics*, 15, pp. 567–585.

Christoffersen, K., and Woods, D. (2002). How to make automated systems team players. In E. Salas (ed.), *Advances in human performance and cognitive engineering research* (Vol. 2, pp. 1–12). Amsterdam: Elsevier Science.

Clancy, J.M., Elliott, G.C., et al. (2003). Command style and team performance in dynamic decision-making tasks. In S.L. Schneider and J. Shanteau (eds), *Emerging perspectives on judgement and decision research* (pp. 586–619). Cambridge: Cambridge University Press.

Clark, H.H., and Brennan, S.E. (1991). Grounding in communication. In L.B. Resnick, J.M. Levine and S.D. Teasley (eds), *Perspectives on socially shared cognition* (pp. 127–149). Washington DC: American Psychological Association.

Conger, J.A., and Pearce, C.L. (2003). A landscape of opportunities: Future research on shared leadership. In J.A. Conger and C.L. Pearce (eds), *Shared leadership: Reframing the hows and whys of leadership* (pp. 285–303). Thousand Oaks: Sage.

Cooke, N.J., and Gorman, J.C. (2009). Interaction-based measures of cognitive systems. *Journal of Cognitive Engineering and Decision Making*, 3, pp. 27–46.

Cox, J.F., Pearce, C.L., and Perry, M. (2003). Toward a model of shared leadership and distributed influence in the innovation process: How shared leadership can influence new product development team dynamics and team effectiveness. In C.L. Pearce and J.A. Conger (eds), *Shared leadership: Reframing the hows and whys of leadership* (pp. 48–76). Thousand Oaks: Sage.

Currall, S.C., and Judge, T.A. (1995). Measuring trust between organizational boundary role persons. *Organizational Behavior and Human Decision Processes*, 64, pp. 151–170.

De Jong, B.A., and Elfring, T. (2010). How does trust affect the performance of ongoing teams? The mediating role of reflexivity, monitoring, and effort. *Academy of Management Journal*, 53, pp. 535–549.

Dirks, K.T. (1999). The effects of interpersonal trust on work group performance. *Journal of Applied Psychology*, 84, pp. 445–455.

Edmondson, A. (1999). Psychological safety and learning behavior in work teams. *Administrative Science Quarterly*, 44(2), pp. 350–383.

Edmondson, A. (2004). Psychological safety, trust, and learning in organizations: A group-level lens. In R.M. Kramer and K.S. Cook (eds), *Trust and distrust in organizations: Dilemmas and approaches* (pp. 239–272). New York: Russell Sage Foundation.

Edmondson, A., Bohmer, R., and Pisano, G. (2001). Speeding up team learning. *Harvard Business Review*, 79, pp. 125–132.

Edmondson, A., and Roloff, K.S. (2009). Overcoming barriers to collaboration: Psychological safety and learning in diverse teams. In E. Salas, G.F. Goodwin and C.S. Burke (eds), *Team effectiveness in complex organizations* (pp. 183–208). New York: Routledge.

Edmunds, A., and Morris, A. (2000). The problem of information overload in business organisations: A review of the literature. *International Journal of Information Management*, 20, pp. 17–28.

Espevik, R., Johnsen, B.H., et al. (2006). Shared mental models and operational effectiveness: Effects on performance and team processes in submarine attack teams. *Military Psychology*, 18(sup1), pp. S23-S36.

Fiore, S.M., Salas, E., et al. (2003). Distributed coordination space: Toward a theory of distributed team process and performance. *Theoretical Issues in Ergonomics Science*, 4, pp. 340–364.

Fiske, S.T., and Neuberg, S.L. (1990). The continuum of impression formation, from category-based to individuating processes: Influences of information and motivation on attention and interpretation. In M.P. Zanna (ed.), *Advances in experimental social psychology* (Vol. 23, pp. 1–74). San Diego: Academic Press.

Flin, R., and Maran, N. (2004). Identifying and training non-technical skills for teams in acute medicine. *Quality and Safety in Health Care*, 13, pp. 80–84.

Gersick, C.J.G. (1988). Time and transition in work teams: Toward a new model of group development. *Academy of Management Journal*, 31, pp. 9–41.

Gersick, C.J.G. (1989). Marking time: Predictable transitions in task groups. *Academy of Management Journal*, 32, pp. 274–309.

Gorman, J.C., Amazeen, P.G., and Cooke, N.J. (2010). Team coordination dynamics. *Nonlinear Dynamics Psychology and Life Sciences*, 14, pp. 265–289.

Gorman, J.C., and Cooke, N.J. (2011). Changes in team cognition after a retention interval: The benefits of mixing it up. *Journal of Experimental Psychology: Applied*, 17, pp. 303–319.

Guastello, S.J. (2010). Nonlinear dynamics of team performance and adaptability in emergency response. *Human Factors*, 52, pp. 162–172.

Hayes, P.A.J. (2012). *Bushfires and other emergencies: Do incident management teams that have worked together make better decisions?* (Unpublished doctoral thesis). La Trobe University: Melbourne.

Hayes, P.A.J., and Omodei, M.M. (2011). Managing emergencies: Key competencies for incident management teams. *The Australian and New Zealand Journal of Organisational Psychology*, 4, pp. 1–10.

Hayes, P.A.J., and Omodei, M. (2012). *Getting the best bang for your buck: Ad hoc or pre-formed Incident Management Teams.* Paper presented at the Third Human Dimensions of Wildland Fire Conference, Seattle.

Hollenbeck, J.R., Ilgen, D.R. et al. (1995). Multilevel theory of team decision making: Decision performance in teams incorporating distributed expertise. *Journal of Applied Psychology,* 80, pp. 292–316.

Huckman, R.S., Staats, B.R., and Upton, D.M. (2009). Team familiarity, role experience, and performance: Evidence from Indian software services. *Management Science,* 55, pp. 85–100.

Ilgen, D.R., Major, D.A. et al. (1995). Raising an individual decision-making model to the team level: A new research model and paradigm. In R. Guzzo, E. Salas and Associates (eds), *Team effectiveness and decision making in organizations* (pp. 113–148). San Francisco: Jossey-Bass.

Kanki, B.G., and Foushee, H.C. (1989). Communication as group process mediator of aircrew performance. *Aviation, Space, and Environmental Medicine,* 60, pp. 402–410.

Katz, R. (1982). The effects of group longevity on project communication and performance. *Administrative Science Quarterly,* 27(1), pp. 81–104.

Katz, R., and Allen, T.J. (1982). Investigating the Not Invented Here (NIH) syndrome: A look at the performance, tenure, and communication patterns of 50 R&D Project Groups. *R&D Management,* 12, pp. 7–20.

Keller, R.T. (1986). Predictors of the performance of project groups in R&D organizations. *Academy of Management Journal,* 29, pp. 715–726.

Kirkman, B.L., Rosen, B., et al. (2006). Enhancing the transfer of computer-assisted training proficiency in geographically distributed teams. *Journal of Applied Psychology,* 91, pp. 706–716.

Klein, G. (1997). The current status of the naturalistic decision framework. In R. Flin, E. Salas, M. Strub and L. Martin (eds), *Decision making under stress: Emerging themes and applications* (pp. 11–28). Aldershot: Ashgate.

Klein, G. (2009). *Streetlights and shadows: Searching for the keys to adaptive decision making.* Cambridge, MA: MIT Press.

Klein, G., Feltovich, P.J., et al. (2005). Common ground and coordination in joint activity. *Organizational simulation* (pp. 139–184). John Wiley & Sons, Inc.

Kozlowski, S.W.J., and Bell, B.S. (2003). Work groups and teams in organizations. In W.C. Borman, D.R. Ilgen and R.J. Klimoski (eds), *Handbook of psychology: Volume 12 industrial and organizational psychology* (pp. 333–375). Hoboken: Wiley.

Kozlowski, S.W.J., and Ilgen, D.R. (2006). Enhancing the effectiveness of work groups and work teams *Psychological Science in the Public Interest,* 7, pp. 77–124.

Kozlowski, S.W.J., Watola, D.J., et al. (2009). Developing adaptive teams: A theory of dynamic team leadership. In E. Salas, G.F. Goodwin and C.S. Burke (eds), *Team effectiveness in complex organizations: Cross-disciplinary perspectives and approaches* (pp. 112–155). New York: Routledge.

Kramer, R.M. (1999). Trust and distrust in organizations: Emerging perspectives, enduring questions. *Annual Review of Psychology*, 50, pp. 569–598.

Leedom, D.K., and Simon, R. (1995). Improving team coordination: A case for behavior-based training. *Military Psychology*, 7, pp. 109–122.

Levine, J.M., Moreland, R., et al. (2005). *Personnel turnover and team performance*. Arlington: United States Army Research Institute for the Behavioral Sciences.

Lewicki, R.J., and Bunker, B.B. (1996). Developing and maintaining trust in work relationships. In R.M. Kramer and T.R. Tyler (eds), *Trust in organizations: Frontiers of theory and research* (pp. 114–139). Thousand Oaks: Sage.

Lewis, A., Hall, T.E., and Black, A. (2011). Career stages in wildland firefighting: Implications for voice in risky situations. *International Journal of Wildland Fire*, 20, pp. 115–124.

Liang, D.W., Moreland, R.L., and Argote, L. (1995). Group versus individual training and group performance: The mediating role of transactive memory. *Personality and Social Psychology Bulletin*, 21, pp. 384–393.

Lipshitz, R., Klein, G., and Carroll, J.S. (2006). Naturalistic decision making and organizational decision making: Exploring the intersections. *Organization Studies*, 27, pp. 917–923.

Lipshitz, R., and Strauss, O. (1997). Coping with uncertainty: A naturalistic decision-making analysis. *Organizational Behavior and Human Decision Processes*, 69, pp. 149–163.

MacMillan, J., Entin, E.E., and Serfaty, D. (2004). Communication overhead: The hidden cost of team cognition. In E. Salas and S.M. Fiore (eds), *Team cognition: Understanding the factors that drive process and performance* (pp. 61–82). Washington, DC: American Psychological Association.

MacMillan, J., Paley, M.J., et al. (2002). Designing the best team for the task: Optimal organizational structures for military missions. In M. McNeese, E. Salas, and M.R. Endsley (eds), *New trends in cooperative activities: System dynamics in complex settings* (pp. 284–299). San Diego: Human Factors and Ergonomics Society Press.

Mailath, G.J., and Postlewaite, A. (1990). Workers versus firms: Bargaining over a firm's value. *The Review of Economic Studies*, 57, pp. 369–380.

McAllister, D.J. (1995). Affect- and cognition-based trust as foundations for interpersonal cooperation in organizations. *Academy of Management Journal*, 38, pp. 24–59.

McEvily, W., Perrone, V., and Zaheer, A. (2003). Trust as an organizing principle. *Organization Science*, 14, pp. 91–103.

McIntyre, R.M., and Salas, E. (1995). Measuring and managing for team performance: Emerging principles from complex environments. In R.A. Guzzo and E. Salas (eds), *Team effectiveness and decision making in organizations* (pp. 9–45). San Francisco: Jossey-Bass Publishers.

McLennan, J., Holgate, A.M., et al. (2006). Decision making effectiveness in wildfire incident management teams. *Journal of Contingencies and Crisis Management*, 14, pp. 27–37.

Meyerson, D., Weick, K.E., and Kramer, R.M. (1996). Swift trust and temporary groups. In R.M. Kramer and T.R. Tyler (eds), *Trust in organizations: Frontiers of theory and research* (pp. 166–195). Thousand Oaks: Sage.

Moreland, R.L., Argote, L., and Krishnan, R. (1998). Training people to work in groups. In R.S. Tindale, L. Heath, et al. (eds), *Theory and research on small groups* (pp. 37–60). New York, Plenum Press.

Morgan, B.B., Glickman, A.S., et al. (1986). *Measurement of team behaviors in a navy environment*. Triangle Park: Battelle Columbus Labs Research.

Nahapiet, J., and Ghoshal, S. (1998). Social capital, intellectual capital, and the organizational advantage. *Academy of Management Review*, 23, pp. 242–266.

Nembhard, I.M., and Edmondson, A.C. (2006). Making it safe: The effects of leader inclusiveness and professional status on psychological safety and improvement efforts in health care teams. *Journal of Organizational Behavior*, 27, pp. 941–966.

NTSB. (1994). *A review of flightcrew-involved, major accidents of US air carriers, 1978 through 1990*. Washington, DC: National Transportation Safety Board.

O'Reilly, C. (1978). The intentional distortion of information in organizational communication: A laboratory and field investigation. *Human Relations*, 31, pp. 173–193.

Omodei, M., and McLennan, J. (2000). Conceptualizing and measuring global interpersonal mistrust--trust. *The Journal of Social Psychology*, 140, pp. 279–294.

Omodei, M., Wearing, A.J., and McLennan, J. (2000). Relative efficacy of an open versus restricted communication structure for command and control decision making: An experimental study. In C. McCann and R. Pigeau (eds), *The human in command: Exploring the modern military experience* (pp. 369–386). New York: Kluwer Academic.

Orasanu, J., and Connolly, T. (1993). The reinvention of decision making. In G. Klein, J. Orasanu, R. Calderwood and C.E. Zsambok (eds), *Decision making in action: Models and methods* (pp. 3–20). Norwood: Ablex.

Orasanu, J., and Salas, E. (1993). Team decision making in complex environments. In G.A. Klein, J. Orasanu, et al. (eds), *Decision making in action: Models and methods* (pp. 327–345). Norwood: Ablex Publishing.

Palmer, E. (1990). *Crew situation awareness*. Paper presented at the Human Factors Society 34th Annual Meeting, Santa Monica.

Pearce, C.L. (2004). The future of leadership: Combining vertical and shared leadership to transform knowledge work. *Academy of Management Executive*, 18, pp. 47–57.

Pearce, C.L., and Conger, J.A. (2003). All those years ago: The historical underpinnings of shared leadership. In C.L. Pearce and J.A. Conger (eds), *Shared leadership: Reframing the hows and whys of leadership* (pp. 1–18). Thousand Oaks: Sage.

Pearce, C.L., and Sims Jr, H.P. (2002). Vertical versus shared leadership as predictors of the effectiveness of change management teams: An examination of aversive, directive, transactional, transformational, and empowering leader behaviors. *Group Dynamics: Theory, Research, and Practice*, 6, pp. 172–197.

Perry, M.L., Pearce, C.L., and Sims Jr, H.P. (1999). Empowered selling teams: How shared leadership can contribute to selling team outcomes. *The Journal of Personal Selling and Sales Management*, 19, pp. 35–51.

Rasker, P.C., Post, W.M., and Schraagen, J.M.C. (2000). Effects of two types of intra-team feedback on developing a shared mental model in command & control teams. *Ergonomics*, 43, pp. 1167–1189.

Reagans, R., Argote, L., and Brooks, D. (2005). Individual experience and experience working together: Predicting learning rates from knowing who knows what and knowing how to work together. *Management Science*, 51, pp. 869–881.

Reid, F.J.M., and Reed, S. (2000). Cognitive entrainment in engineering design teams. *Small Group Research*, 31, pp. 354–382.

Rempel, J.K., Holmes, J.G., and Zanna, M.P. (1985). Trust in close relationships. *Journal of Personality & Social Psychology*, 49, pp. 95–112.

Rousseau, D., Sitkin, S.B., et al. (1998). Not so different after all: A cross discipline view of trust. *Academy of Management Review*, 23, pp. 393–404.

Salas, E., Dickinson, T.L., et al. (1992). Toward an understanding of team performance and training. In R.W. Swezey and E. Salas (eds), *Teams: Their training and performance* (pp. 3–29). Norwood: Ablex.

Salas, E., Guthrie, J.W., et al. (2005). Modelling team performance: The basic ingredients and research needs. In W.B. Rouse and K.R. Boff (eds), *Organizational simulation* (pp. 185–228). Hoboken: Wiley-Interscience.

Salas, E., Rosen, M.A., et al. (2009). The wisdom of collectives in organizations: An update of the teamwork competencies. In E. Salas, G.F. Goodwin and C.S. Burke (eds), *Team effectiveness in complex organizations: Cross disciplinary perspectives and approaches* (pp. 39–79). New York: Routledge.

Schmitt, J.F., and Klein, G.A. (1996). Fighting in the fog: Dealing with battlefield uncertainty. *Marine Corps Gazette*, 80, pp. 62–69.

Schraagen, J.M., and van de Ven, J. (2011). Human factors aspects of ICT for crisis management. *Cognition, Technology & Work*, 13, pp. 175–187.

Simon, H.A. (1955). A behavioral model of rational choice. *The Quarterly Journal of Economics*, 69, pp. 99–118

Simon, H.A. (1979). Rational decision making in business organizations. *The American Economic Review*, 69, pp. 493–513.

Steiner, I.D. (1972). *Group process and productivity*. New York: Academic Press.

Tuckman, B.W. (1965). Developmental sequence in small groups. *Psychological Bulletin*, 63, pp. 384–399.

Weick, K.E., and Roberts, K.H. (1993). Collective mind in an organizations: Heedful interrelating on flight decks. *Administrative Science Quarterly*, 38, pp. 357–381.

Wildman, J.L., Shuffler, M.L., et al. (2012). Trust development in swift starting action teams. *Group & Organization Management,* 37, pp. 137–170.

Wilson, J.M., Straus, S.G., and McEvily, W. (2006). All in due time: The development of trust in computer-mediated and face-to-face teams. *Organizational Behavior and Human Decision Processes,* 99, pp. 16–33.

Wittenbaum, G.M., Vaughan, S. I., and Stasser, G. (1998). Coordination in task-performing groups. In R.S. Tindale, L. Heath, et al. (eds), *Theory and research on small groups* (Vol. 4, pp. 177–204). New York: Plenum Press.

Zaheer, A., McEvily, W., and Perrone, V. (1998). Does trust matter? Exploring the effects of interorganizational and interpersonal trust on performance. *Organization Science,* 9, pp. 141–159.

Zalesny, M.D., Salas, E., and Prince, C. (1995). Conceptual and measurement issues in coordination: Implications for team behaviour and performance. In G.R. Ferris (ed.), *Research in personnel and human resource management* (Vol. 13, pp. 81–115). Greenwich: JAI.

Chapter 7

Leadership, Communication and Teamwork in Emergency Management

Dr Christine Owen

Bushfire Cooperative Research Centre, University of Tasmania, Australia

Introduction

Effective communication within and between teams is increasingly recognised as an important component for adaptability to changing conditions (Burke, Stagl, Salas, Pierce and Kendall 2006; Crichton, Lauche and Flin 2005; Flin, O'Connor and Crighton 2008). As has been discussed in other chapters, there are a wide variety of teams in use in emergency management. This is because emergency events typically involve different groups of people coming together to work on their part of the impact. These can include specialist teams (e.g., fire behaviour modelling teams, or municipal recovery teams). Within incident management teams (IMTs) there are functional unit sub-teams (e.g., operations, planning, logistics; see Chapter 1, this volume). All these teams need to interact with teams in other locations and organisations (see Chapter 1, this volume). Teams are also becoming increasingly specialised (Nembhard and Edmondson 2006). They need to interact with other specialist teams when addressing complex and multifaceted problems (Fiore 2008; Zaccaro, Ardison and Orvis 2004). These complexities increase the need for coordination between interdisciplinary and specialist teams leading to increased interdependence between these teams (DeChurch and Zaccaro 2010; Zaccaro, Heinen and Shuffler 2009).

This interdependency is also complex because it involves reciprocal (as well as sequential) interactions (Nembhard and Edmondson 2006). Members of various emergency management teams simply cannot do their jobs and assume others will come along at some point to do theirs. Instead their knowledge and efforts must be integrated to deliver effective emergency management outcomes.

Given the open-ended and unbounded nature of emergency events, these trends – that is, the increasing needs for adaptability, specialisation and interdependence – are more prominent in the emergency management sector than in other safety-critical industries. Together they imply a need for a fresh approach to examining effective communication within teams as well as the collaboration needed across teams from different organisations and disciplines.

Barriers to Information Flow in Emergency Management

Despite the need for collaboration within and between emergency management teams, there are many challenges that must be overcome.

The significance of communication between groups and organisations involved in emergency events is highlighted by how frequently there are breakdowns in both the coordination of the emergency response and the subsequent secondary impacts that occur when coordination has failed. For example, there are a number of reports in the literature (see Comfort and Kapucu 2006; Wise 2006; Scholtens 2008), as well as post-event inquiries both in Australia (e.g., Teague, McLeod and Pascoe 2010) and internationally (e.g., Reid 2006; Moynihan 2007; Lutz and Lindell 2008) that have identified problems in team communication as well as the need for improvements to between-team information flow.

In the USA, for example, information flow between the layers of emergency management was criticised in the review of the response to Hurricane Katrina (Wise 2006). Parochial turf boundaries saw more lives lost than might have been necessary in the collapse of the two towers at the World Trade Center in September 2001. Marcus, Dorn and Henderson (2006) noted that historic rivalry between the police force and the New York Fire Department inhibited information sharing between the two groups, with drastic consequences. The police, from the vantage point of their helicopters, could see that the towers were about to collapse and conveyed this to their people on the ground, but this information was not shared with the firefighters within the building. In Australia cultural and structural impediments to information sharing between two incident management teams was blamed, in part, for the tragic deaths of many in the Victorian 2009 bushfires. Thus, one of the challenges in emergency management is to create effective communication pathways for functional units within teams as well as between teams that may be in distributed locations. Overcoming communication barriers within and between teams is therefore of critical importance.

Communication Coordination and the Role of Leadership

Leadership has been identified as a key variable in the functioning of a team's success or breakdown (e.g., Schauboeck, Lam and Cha 2007; Sy, Cote and Saavedra 2005; Taggar and Seijts 2003). Leaders serve as coordinators of operations, as liaisons to external teams or management (e.g., Burke et al. 2006; Zaccaro, Heinen and Shuffler 2009). Effective leaders encourage collective goals and objectives to define a vision (e.g., Conger and Kanungo 1998; Crichton, Lauche and Flin 2005), generate and maintain confidence, cooperation and trust within the team (e.g., Vogus and Sutcliffe 2007). They also model constructive thinking ability to solve problems (e.g., Katz and Epstein 1991; Salas, Rosen, Burke, Goodwin and Fiore 2006) and are able to identify the associations amongst confronting multiple issues so they can be dealt with concurrently (e.g., Yukl and Van Fleet 1992; Flin, O'Connor and Crichton 2008).

While there is an array of research on the influences leaders have on team effectiveness, minimal attention been given to the role of leaders as team mentors or coaches, particularly in overcoming barriers in communication and coordination. Team coaching is defined by Hackman and Wageman (2005) as 'direct interaction with a team intended to help members make coordinated and task-appropriate use of their collective resources in accomplishing the team's work' (p. 269).

Other studies have shown that communication patterns and climate are influenced by the Team Leader (Nembhard and Edmondson 2006). Team members are highly sensitive to the behaviour of leaders and examine leader actions for insight about what is acceptable and expected in communication (Carson, Tesluk and Marrone 2007; Edmondson 2005; Hambley, O'Neill and Kline 2007). Nembhard and Edmondson (2006) found that Team Leader invitation to share information and appreciation of the contribution was important for facilitating psychological safety.

In an earlier study Edmondson (2003) explored leadership, context and team processes with interdisciplinary action teams. Action teams were 'teams in which members with specialized skills must improvise and coordinate actions in intense, unpredictable situations' (Sundstrom et al., cited in Edmondson 2003, p. 1421). Edmondson found that communication and coordination outside the team with other hospitals and specialists was strongly associated with successful implementation of new practices. In addition the study showed that, in part, leaders' attitudes towards such external communication showed others it was important to have frequent contact with external stakeholders. There was also a significant relationship between Team Leader coaching and a team climate which allowed for team members to have the opportunity to speak up and share what they knew to improve effectiveness.

From this brief review of the literature the following can be concluded:

- how a team performs is in large part influenced by the role the Team Leader;
- attention to external communication is important, particularly for teams engaged in intense, unpredictable situations where improvisation is necessary;
- the Team Leader's attitude to internal and external communication practices influences team climate and team member capacity to speak up and express concerns and needs.

However, none of the established constructs – coaching or psychological safety – sufficiently captures the ways in which Team Leaders may facilitate the information seeking and giving that establishes effective within- and between-team communication and coordination. This then is the contribution of this study.

Method

The aim is to contribute knowledge of the role of the leader in facilitating information sharing and seeking within hierarchical teams and between teams needing to coordinate across disciplinary boundaries. Research questions include:

1. What is the relationship between incident management Team Leadership behaviour and team performance?
2. What communicative practices do incident management Team Leaders use to enhance teamwork performance?
3. How do incident management Team Leaders facilitate team member communication within and between teams?

The Training

The research reported here is based on observations conducted of 18 IMTs undertaking training exercise simulations in five states of Australia. In Fire and Emergency Management in Australia all IMT personnel are required to participate in training simulation as part of the accreditation and refresher process. The training exercises observed were all aimed at qualifying and/or refreshing personnel who were deemed ready and able to manage a Level 3 emergency incident (i.e., the highest level of incident complexity, requiring delegation of all IMT functions – see Chapter 1, this volume). The training was nationally accredited and conducted by experienced accredited trainers.

The Simulation Exercise

The simulation exercises observed all presented intellectually demanding tasks, in real-time[1] to match the conditions of wildfires, and all were between four and eight hours in duration. A wildfire is reported and an IMT has been established. During the course of the exercise the fire escalates and threatens local assets (e.g., a forest plantation and a town). During the simulation there are other problems encountered that need to be managed (these could include injured firefighters; unannounced arrival of political leaders; self-evacuation of community members; spurious media reports).

All the exercises observed were conducted in either actual incident control centres in local fire stations or at sites where IMTs would normally be expected to set up. Thus participants employed the communication processes they would normally use if managing an incident (e.g., radio, telephone, facsimile and paper-based reporting templates).

---

1   i.e., if a requested piece of equipment was 20 minutes away, then it took 20 minutes before it was available to the exercise participants.

The Participants

The participants with key roles (e.g., Incident Controller, Operation Officers, Planning Officers and Logistics Officers) all had considerable experience in firefighting and the majority had previous experience of being involved in an IMT. All had previous desktop training in their IMT role or function and were assumed to already be considered competent for their IMT role or have the capacity to undertake the role allocated.

Ninety per cent of controllers had more than 15 years' experience (65 per cent had more than 20 years). In some of the exercises there were some team members with little or no experience of working in an IMT, though the median years of experience for team members was 11. Team members also had a median 11 years' experience in the industry, although there was a wider range of incident management experience, with some having no experience and others having decades worth.

After the training, participants were asked to assess how close to reality the problems that were presented in the scenario were. Of the Incident Management Team participants included in this study, 99 per cent thought the exercise scenario was like or very like a real fire experience (58.3 per cent of the Incident Management Team participants thought that the exercise was very like a real fire). Team size varied between 15 and 30 people (median = 23) which is typical of a Level 3 IMT.

The Measures

The research design employed a mixed methods approach that included subjective measures of leadership effectiveness by team members as well as independent assessments by the subject matter experts responsible for the training design. In addition, the data collected included talk of the selected Team Leaders for analysis of communication and teamwork behaviour.

Perceptions of Teamwork and Team Leader Effectiveness

During the observations pre- and post-exercise self-report measures of perceived communication and teamwork effectiveness were undertaken. The items were based on team effectiveness indicators developed in other high-consequence domains (e.g., Cannon-Bowers and Salas 1997, 1998; Langan-Fox, Anglim and Wilson 2004; Mohammed and Dumville 2001; Salas, Rosen, Burke and Nicholson 2007; Salas et al. 2006; Mathieu, Maynard, Rapp and Gilson 2008; Volpe, Cannon-Bowers and Salas 1996). Agreement with each statement was measured on a seven-point Likert scale from 1 (which was equal to 'not at all') through to 7 (equal to a 'very great extent'). Participants were also given an option to answer 'don't know'. Participants provided their own unique code to both pre- and post-exercise surveys, allowing pre- and post-subjective measures to be individually matched. Difference scores for each person were calculated, which were then collated to give an overall group score. In addition there were three items measuring perceived leader effectiveness:

- 'The Incident Controller will invite [did invite] input from other team members';
- 'The Incident Controller will make [did make] good decisions';
- 'The Incident Controller will be clear [was clear] in stating goals and intentions'.

The scale measure for the three items to measure team member perceptions of Team Leader effectiveness yielded a Cronbach Alpha scale reliability of α = .82. In addition to the grouped team member leadership effectiveness scores, the accredited instructors were also asked to rate, on a scale of 1–10, their (a) level of satisfaction with the overall team performance and (b) level of satisfaction with the Incident Controller performance. These indicators of perceived Incident Controller effectiveness (the grouped scores on the three team member items plus the two scores provided by the accredited trainers) provided an overall measure of perceived Incident Controller effectiveness. The 18 Incident Controllers were thus ranked based on this composite score.

Audio Transcripts

Audio recordings were taken for key personnel, including the Incident Controllers, by using lapel microphones. For the Incident Controllers a two-hour sample was taken from four time periods within the exercise and transcribed for further analysis. The four time periods were chosen based on accredited trainer advice to represent low tempo (getting established, simple problems) and high tempo (escalating incident, complex problems) simulation periods. Incident Controller talk was then coded against a framework that included:

- direction of information exchange (target initiates or responds)
- information flow (within the IMT or between the IMT and other functions within the ICS)
- types of communicative practices (factual statements; action planning statements; uncertainty; instructions/giving directions; statements for the purpose of providing situation awareness; incomplete statements; banter/ humour)
- supportive teamwork practices observed (offering, requesting, receiving assistance; monitoring/crosschecking; negotiation; team feedback/coaching).

The focus in this chapter is on the supportive teamwork behaviours observed and in particular the ways in which Team Leaders engaged in team feedback/coaching.

Triangulation of Qualitative Coding

Two coders (the author and a research assistant) worked through five pages of transcript, and intra-class correlation coefficients were calculated. The reliability ranged from .75 to .80. Following discussion a new five-page transcript was coded and the inter-rater reliability was .88. Once this required level of reliability was achieved all transcripts were then coded.

The scores from the perceived leadership effectiveness variable were ranked and the top 25 per cent included five Incident Controllers and their teams (five teams comprising 95 members); and the bottom 25 per cent included five Incident Controllers (five teams comprising 88 members).

Based on this identification the audio data collected on Incident Controller talk was combined into two groups: those perceived as being more effective ('effective Team Leaders') and those who were perceived as less effective ('less effective Team Leaders'). Having undertaken this compilation of the data it was then possible to analyse the talk in terms of whether the teamwork behaviours were different between the two groups. Examples of the indications of teamwork behaviour are summarised in Table 7.1.

For the purposes of this chapter the utterance or interaction coded is the unit of analysis. It can be seen from Table 7.2 that collectively, when the Team Leaders were observed, they were engaged in a total of 378 (i.e., totals of 209 and 167) teamwork supporting behaviours.

Table 7.1   Definitions of teamwork behaviours

Teamwork Behaviours	Supporting Behaviour, Team Communication
1 Offering or providing assistance	Ask if a team member would like a hand or initiating support without being asked
2 Receiving	Graciously accepting assistance offered
3 Requesting assistance	Seeking help when needed
4 Monitoring	Checking on the progress or well-being of a team member – seeking information on how they are going
5 Negotiation	Requiring agreement or endorsement of action statement, planning statement, request something that needs to be sorted or understood
6 Flexibility	The ability and willingness to adapt performance strategies quickly and appropriately to changing task and demands
7 Team feedback coaching	Team management, facilitating teamwork. Team members are invited/coached to communicate their observations, concerns, suggestions and requests in a clear, direct and assertive manner
8 Banter/joking/sarcasm	Team members bantering, joking or speaking with sarcasm

Table 7.2   Observed teamwork behaviours

	Less Effective Team Leaders	Effective Team Leaders
Offering or providing assistance	26 (15%)	23 (11%)
Receiving assistance	35 (21%)	4 (2%)
Requesting information or assistance	16 (9%)	33 (16%)
Monitoring	38 (22%)	45 (21%)
Negotiation	5 (3%)	18 (8%)
Team feedback/coaching	41 (24%)	81 (39%)
Banter/joking/sarcasm	8 (5%)	5 (2%)
TOTAL	**169**	**209**

How Leaders Support their Teams

Table 7.2 shows the results of the number of observed teamwork supporting behaviours coded to the two groups. It can be seen that the number of observed teamwork behaviours are lower for the perceived less effective group; but not dramatically so – 169 observed supportive behaviours (45 per cent of all supported teamwork behaviours observed) for the less effective leader group compared to 209 (or 55 per cent) for the perceived effective leader group. It can be anticipated that the quality of the statements differs and this was examined for one particular category worthy of note. What is of interest is the difference in the amount of team behaviour that was observed and coded as team feedback. Team feedback behaviours were those where the Team Leaders provided information to team members on what they needed to do or to be thinking about. Team members were also coached to communicate to others their observations, concerns, suggestions and requests in a clear, direct and assertive manner. For example: '*Can you make sure you're talking to XX in Resources and let them know…*'

The more effective Team Leader group provided more frequent feedback and coaching communication than those performing less well (effective Team Leaders n = 81; less effective Team Leaders n = 41; see Figure 7.1). Given the important role that Team Leaders play in facilitating and coaching others to communicate their needs and insights, the interactions indicating supportive team feedback were extracted for further examination.

A qualitative analysis was then performed on these 122 feedback and coaching communication extracts to look for themes. As would be expected, in addition to the difference in quantity there were also differences in the focus indicated in the extracts and it will be argued that the quality of communication was also different between the two groups.

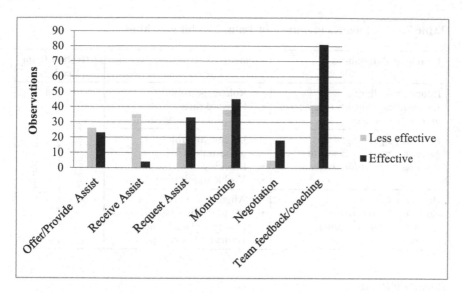

Figure 7.1   Number of observations of supportive teamwork behaviours for effective and less effective Team Leaders

Team Leader Coaching

The thematic analysis of the extracts indicated that Team Leaders perform three coaching roles to facilitate communication and coordination within and between teams. Team Leader feedback was directed towards coaching team members to address three broad purposes:

- the temporal demands of the task and making adjustments needed to changing conditions (i.e., riding the boundaries of the team)
- to integrate their activities within the team (i.e., spanning the boundaries within the team)
- reporting and seeking cooperation with others outside the team (i.e., crossing the boundaries between teams).

These three purposes have been defined as Boundary Riding, Boundary Spanning and Boundary Crossing coaching behaviours. The themes and their elements are outlined in Table 7.3.

Table 7.3   Themes identified in Team Leader coaching

Coaching Behaviours	Elements	Effective Team Leaders	
Boundary Riding: *Ensuring adjustments to temporal and event demands*	• Role responsibility • Forward focus • Transition management	Overlapping-ness	With-it-ness
Boundary Spanning: *Ensuring integration within function units within the team*	• Linking and role orchestration • Blocking • Visible monitoring		
Boundary Crossing: *Ensuring alignment and coordination across team boundaries*	• Aligning expectancies • Strengthening assertiveness • Managing lateral and vertical relationships		

Boundary Riding

Coaching for Boundary Riding included activities designed to assist team members to clarify and exercise their own responsibilities and to be mindful of the impact of these responsibilities on others. This also involved coaching team members in their responsibilities to contribute to team thinking and problem solving. The emphasis was on constantly calibrating and adjusting to the temporal demands in managing the event and to any perceived changes. This involved three sub-themes pertaining to role responsibility, to maintaining a forward focus and to effectively managing transitions needed.

Role responsibility

Team Leaders assisted team members to be clear on their own roles as well as how those role responsibilities related to others. Effective Team Leaders would ensure team members were clear on what they needed to be doing, as well as what they need to be looking out for that signalled a need to change.

Part of the responsibility of team members is to make any concerns known and to contribute thinking to problem solving. Rather than a generic invitation made by a less effective Team Leader *'raise that at the meeting would you?'*, effective Team Leaders were specific in their expectations that others should have something to add:

> So, I'd like some input from everybody on what I've covered – very quickly –
> any amendments or improvements to that I'd really welcome. If I could start
> with you [name] in the planning division.

Role responsibilities within the team also included an expectation of team member responsibilities towards other team members:

> I think you also have responsibilities in your part of the operations. If they need
> a hand as it gets busier – go and give them a hand.

In the above quote above the Team Leader is setting the tone for the climate and establishing norms for offering and providing assistance to others. By stating this as part of the leader's expectations these supporting behaviours can become normalised within the team. Effective Team Leaders were also explicit in noting the obligation team members have to look out for one another and to intervene when needed:

> Just check that they are having breaks. Because they're starting to get bit a busy
> and stressed. Make sure they have a cup of coffee and a break and have got relief
> amongst themselves.

Forward focus

Boundary Riding also included emphasising the temporal demands of the incident and its implications so that all team members could fulfil their responsibilities. In this respect Team Leaders emphasised a *forward focus*. While leaders who were perceived as less effective might make a general comment about future conditions, more effective Team Leaders were specific about what to look for and why it matters. While a less effective Team Leader might say something rather generic such as:

> I want everyone to be proactive, look as far ahead as we can at our main risks
> and, ah, get ahead of it.

In contrast, a more effective leader was specific about what needed to be the focus of concern and why:

> The way I read this is if we don't get control of this fire within the next 24 to 48
> hours at the latest, then we've got a major campaign fire and as you're aware
> we've got about three risk groups that are threatened [names them]. We've got
> dry air and dehydration is also a safety issue. It might get a run on us so we need
> to regular updates of…So let's set up some time-frames on the white board and
> test them to see how realistic they are.

The forward focus also included the Team Leader identifying potential problem areas and nipping these 'in the bud' before they became a problem. The following Team Leader draws attention to the responsibilities of a team member and assures good practice:

> Excuse me [name]? Just another little thing which you probably already know
> but I just thought I'd mention. If you could have one of your staff check

everybody coming in has their safety gear and that they get a safety briefing when they arrive. Just a quick reminder, a quick refresh would be a really good thing. And obviously, there's a need to track people's times. Thank you.

In this quote the Team Leader is letting the team member know that they need to address a potential oversight in their responsibilities, which includes delegating others to complete particular tasks. The quote also suggests that there is a quick way of doing this, which again coaches about the importance of efficient communication in the team. The chunking of 3–4 main elements also indicates efficiency in conveying these expectations rarely seen in statements made by less effective Team Leaders.

Transition management

Effective Team Leaders were also particularly efficient in their own transition management and in coaching this in others. This may involve keeping up to date themselves:

> How long do you need? 4 or 5 minutes? Actually if you've got a copy, I'll just have a quick look at it now.

This kind of practice enables the Team Leader to gain a quick appreciation of the situation, provides an opportunity to make any corrections needed and supports the team member in feeling confident they are on the right track. Effective Team Leaders also engaged in managing multiple overlapping tasks in a way that was not too disruptive but supported efficiency in momentum:

> As you know, the first thing I've got to get away is the letter of appointment so you'll all have to read and sign this so, ah, I'll just pass this form round as we start the briefing to get it out of our road and then we'll get logistics to fax that off to the [place].

Effective Team Leaders also drew attention to where team attention was focussed, which in some cases highlighted a need to switch gears:

> …at the moment people are very focussed on operational fire fighting; you need to start thinking about logistics. I want you to look after the team.

In this respect the Team Leader is verbalising a metacognitive thinking process (see Frye and Wearing, Chapter 4, this volume), drawing attention to a need to effect a transition in activity. Doing so makes the narrowing of focus (onto operational firefighting activities) visible and thus shared so that other team members become aware that there may be a risk of attention narrowing. This type of observation also supports mindfulness because it draws attention to the need to proactively manage the risk of a narrowing attention focus and losing sight of other needed activities. In contrast, leaders perceived as less effective seemed not to notice a narrowing of attention in others and indeed sometimes contributed to tunnel vision, drawing other team members along with them.

While Boundary Riding principally involved the Team Leader engaging in information seeking and information giving to address temporal demands and adjustments, in Boundary Spanning and Boundary Crossing, the Team Leader coaches others in information seeking and giving for particular purposes.

Boundary Spanning

The intention with Boundary Spanning is to ensure that there is effective integration within and between functional units within the team. In so doing there was an emphasis on updating inter-positional knowledge. Having inter-positional knowledge (i.e., knowledge about the environment, tasks, roles and appropriate behavioural responses required of team mates in various situations) assists with the development of shared mental models (Volpe, Cannon-Bowers and Salas 1996). Coaching for functional integration between member roles and responsibilities occurred in a number of ways.

Role orchestration linking

Linking is when the Team Leader is coaching team members to work together in their information sharing activities. For example:

> To get those regular updates, we're going to meet with [name] on the hour every hour and you and [name] need to be hand and glove.

> I'm concerned they're [Planning] not getting regular updates.

Boundary Spanning also emphasised the need for temporal synchronisation to achieve team goals to gain full efficiency in reciprocal coordination:

> Just be mindful you're going to have to give those people a bit of notice.

Such coaching consisted of both communications needed to seek information from others and then to share that information with others:

> Well, you need to get onto [name] to find out what they're doing and how they're going to run it. And then you need to pass that back to Planning.

> As soon as we get feedback from resources branch of what's coming, you'll need to let Logistics know so Logistics can make sure they have water and fuel when they get in.

This kind of coaching supports team members to attend mindfully to interconnecting activities. It enables the team to function with greater efficiency, particularly in terms of addressing the temporal demands and the needed synchronisation between task responsibilities of functional units.

Blocking

In additional to linking through role orchestration, Team Leaders engaged in blocking activity. This was designed to short-circuit information coming to them that needed to be given to someone else. In the case of less effective Team Leaders, in some cases they received the information and then took the time to directly convey the information themselves to the intended recipient. In contrast, effective Team Leaders would cut off the information giver and redirect, as indicated in the following comments:

> Yep, that's already being undertaken by the deputy so if you can just raise that with the deputy first.

> Look I'd like you to go and speak to Jason here. Jason's our operations person. Um, Jason would also probably like to have a chat to you [too] and get an update and, ah, on what's been happening.

The above quote reminds both parties of their obligations to one another. The Team Leader establishes the expectation that once the operations officer addresses the information-seeker's concerns, he/she should expect some update since the seeker probably has valuable information that the Operations Officer can use. Once again these types of interlinking activities offer efficiencies not in evidence in the talk of less effective Team Leaders.

Visible monitoring

Visible monitoring involved making the observations made by the Team Leader also visible to others. The sub-theme of visible monitoring ensures the Team Leader lets others know about disconnects identified between people and/or functional units to alert others (e.g., functional unit leaders) that they need to intervene if necessary to redress potential breakdowns. For example:

> I've just spotted an issue between Logs and Ops. They thought they'd organised some relief crews and thought they were under way when in fact they weren't.

In this respect the Team Leader has observed an oversight and passes this information on. In some respects this activity required coaching functional unit leaders in how to intervene rather than the Team Leader doing it themselves.

Rather than a vague comment made by a less effective Team Leader such as '*there seems to be a problem out there, can you sort it out?*', effective Team Leaders were much more proactive, highlighted potential problems earlier and were more specific, sometimes to the point of being directive:

> I suspect that Jake could use some help on the mapping as that doesn't seem to be progressing. Could you find someone to take on that role?

In this respect they appeared to be more observant of what was going on with team members, identified potential sources of problems before they became too troublesome, and proactively took steps to remediate any potential communication or coordination issues.

Sometimes visible monitoring overlapped with the next theme, Boundary Crossing. This can be seen in the next quote where a Team Leader uses the time to have a conversation with someone in transit about a particular issue and also pointedly notes a need to keep others informed:

> Ah, good morning [name], it's [name], Incident Controller at the fire here at [name of town],…Um, I believe you're going to get into [name of town] in another 15 minutes or so. Ah, 60 minutes, ok, thank you for confirming that. I'd better let Logistics know that. While you're driving up, if I can quickly chat to you.…

In this respect there is a gentle chiding of a need to update and inform others. This is something that is not only important within functional units but also with members of other teams – which is the purpose of Boundary Crossing activities.

Boundary Crossing

In Boundary Crossing the focus is on meeting the information needs and concerns of other teams operating within the emergency management arrangements and other external stakeholders. Effective Team Leaders did this by working to align expectancies between the IMT and other teams or stakeholders.

Aligning expectancies

Effective Team Leaders were frequently directly engaged in Boundary Crossing activities – more so than those perceived as less effective (see Table 7.4). Team Leaders conferred with external stakeholders regarding the expectations and requirements for both information seeking and information giving.

This Boundary Crossing activity is then used to reinforce Boundary Riding as well as Boundary Spanning activities. For example, in the following quote the Team Leader uses the information gained to ensure team members are mindful of external stakeholder concerns:

> They were a bit concerned that our fire must not leave [name of place].

Sometimes addressing the need to cross boundaries to establish linkages with external teams requires Team Leaders to task team members to take on Boundary Crossing responsibilities, as in the following two quotes:

> If we get any more [name of agency] action happening we may need to put someone down as a [name of agency] liaison.

> Quite honestly from where I sit, ah, we have a major communication issue, um, particularly the stakeholders around this fire, geographically situated around this fire. The stakeholders are obviously the Council. But also the [critical infrastructure], Police, and we need to also get through to a lot of the ecotourism resorts in the area. So we need a sort of like a pole transition if you like. Do you think you can manage to make those contacts and feed them back to [team member name]?

In the above quote the Team Leader is not only expressing his concerns but also coaching the listener to be thinking about them and encouraging that team member to become a conduit for information between the town and a member of the IMT.

Boundary Crossing also connects to Team Leader Boundary Riding activities intended to circumscribe or limit roles and responsibilities and to be mindful of external team knowledge and legitimacy:

> Ask him. Its, we'll need to get onto them. We don't have any expertise in buildings so it's a real [name of agency] job.

Strengthening assertiveness

Team Leaders also coach other team members to be conducting their own Boundary Crossing activities. This requires team members to be outwardly focussed and to be also clear in managing external stakeholder expectations. It sometimes requires leaders to coach team members in strengthening their assertiveness:

> Well, my background would indicate that if you put coppers up there and say no-one is to come through [place]. If they close the road, they close the road. They won't let anyone through. So that really needs to be spelt out to them that local traffic, residents or owners have access but it's not open to general traffic. Can you check, [name], that she is very clear with the MERC. Say that 'there should be no one running around saying...'

In this respect messages need to be made clear and suitably forceful, which requires coaching, particularly if team members are speaking to stakeholders they typically don't work with all the time or where there may be a difference in power.

Relationship management

In addition, coaching is provided to ensure team members conduct Boundary Crossing activities efficiently:

> You're way better off to – even if it's the local cop, or a local member…all that stuff, do it in front of the map. You can point it out. That's why that map is so important.

In the above quote the coaching in the use of the map is to help convey to external stakeholders the key messages as effectively and as efficiently as possible.

Overlapping-ness and With-it-ness[2]

In all types of coaching activity Team Leaders perceived as more effective also had greater clarity and depth in their statements. They engaged in management practices known as 'overlapping-ness' and 'with-it-ness'. In communicating their expectations they frequently demonstrated and coached team members on multi-tasking (overlapping-ness). In addition they appeared more engaged and aware of what was happening (with-it-ness).

As the examples used earlier in the chapter indicate, effective Team Leaders frequently chunked key messages with 3–4 activities and gained improved efficiencies. In contrast, less effective Team Leaders frequently just focussed on one thing at a time. Effective Team Leaders also overlapped these chunks, sometimes engaging in multiple (e.g., Boundary Spanning and Crossing activities). This is indicated in Table 7.4 by the number of interactions that started with one type of coaching purpose and included a secondary purpose. Where a comment included two types of coaching (i.e., where the statement might start with Boundary Riding coaching and transition into either Boundary Spanning or Boundary Crossing), this is represented in brackets. The table shows that effective Team Leaders engaged in 21 Boundary Crossing interactions, of which seven overlapped to also include either riding or spanning purposes. In this way effective Team Leaders demonstrated greater proficiency in making their communications count. They combined the management of role responsibilities, within-team integration and between-team relationship and expectations management.

---

2 With due deference to Jacob Kounin *Discipline and Group Management in Classrooms*. Huntington, NY: R.E. Krieger, 1977, c. 1970.

Table 7.4   Types of coaching undertaken by Team Leaders

	Less Effective Team Leaders	More Effective Team Leaders
Boundary Riding	24	47 (3)
Boundary Spanning	7	34 (14)
Boundary Crossing	9	21 (7)
	40	**102**

Note: ( ) indicates number that overlapped with another coaching category

Effective Team Leaders also displayed more 'with-it-ness'. As indicated in the quotes, they appeared to be more aware of what was happening and what steps were needed to address any deficiencies. They also extended their reach of influence beyond those with whom they were directly interacting. In many of the Boundary Spanning and Boundary Crossing coaching activities there was attention not only to directly engaging the person being spoken to but also to suggesting what they might do to connect/remind/obtain action from others. In the following example the Team Leader is encouraging the listener (second party) to contact a third party to get action from a fourth:

> [Name], can you ask [name] to ring the RO please. And tell him to talk to the DERC. Tell him that there is a school camp.

In this respect Team Leaders are extending their own sphere beyond those who they might be able to directly influence to attempt to obtain cooperation from other stakeholders of concern. This is in contrast to most comments from less effective Team Leaders who tended to be focussed on the actions of the person to whom they were directly speaking.

Similarly in the following quote there is concern to let stakeholders know about the anticipated secondary impacts:

> If [name] can brief Planning, and [name] can talk to the MERC and let him know that these are the roads that we're looking at. And this is where we think it's going to go.

These examples illustrate the means by which effective Team Leaders were both more efficient in their coaching and more focussed.

Implications for Practitioners and Instructors

Organisations deliver highly reliable performances when members have the ability to prevent and manage mishaps before they spread throughout the system causing widespread damage or failure (Barton and Sutcliffe 2009, p. 1329). This occurs when team members engage in social mechanisms for monitoring and reporting small or weak signals (i.e., that something might be wrong) to one another to provide a shared situation awareness necessary to enable team members to adjust to these changing conditions. Thus members have both the flexibility required and the capability to respond in real-time, reorganising resources and actions as necessary. From this perspective then, high reliability organising and safety is achieved through human processes and relationships (Barton and Sutcliffe 2009). Members share and seek information with each other about what they know or are concerned about and the team adjusts, tweaks and adapts to this information flow which, if not forthcoming, could result in larger problems and potential failures in safety. This chapter has demonstrated the ways in which Team Leaders coach members to engage in these communications processes and relationships.

Facilitators of learning in emergency management teamwork will be able to use these insights in a variety of ways. First, the practices of Boundary Riding, Boundary Spanning and Boundary Crossing can be discussed and modelled in leadership development programmes as well as in exercise simulations. Discussing the presence or absence of these activities can be included in exercise debriefs. In addition, the material provided here could be further developed into a checklist for instructor feedback to leaders, as well as for use in real-time performance monitoring. In this respect it can complement the ideas of facilitating non-technical skills (see Brooks, Chapter 9, this volume) through enhancing metacognitive thinking (see Frye and Wearing, Chapter 4, this volume) and critical reflection (see Stack, Chapter 10, this volume).

Leadership, communication and teamwork in emergency management have come under intense scrutiny world-wide. Teams in complex environments rely heavily on their leaders to continually develop and promote a shared understanding of the situation and problems that arise, and what resources (including members' capabilities) might be called upon to ensure safe and effective performance. As teamwork becomes more intense and complex, and as multiple team coordination becomes more interdependent, it is crucial that leaders and their teams are able to manage these dynamic conditions. This chapter contributes to identifying how effective Team Leaders coach their teams under these conditions and in so doing assist others to be able to develop Boundary Riding, Spanning and Crossing capabilities.

Conclusion

The purpose of this chapter has been to examine the ways in which Team Leaders coach team members to enhance within- and between-team communications to enable effective management of emergency events. While it has been widely demonstrated that the way in which teams work is critical to effective emergency management, it is also important not to forget that emergency management is more commonly a problem of communication between teams and between organisations. Moreover, there is evidence emerging to suggest that it is not effective teamwork *per se* that makes the difference in complex multi-layered multi-team systems, but is how the boundaries between teams are managed that is important (De Church and Zacarro 2010). In this respect effective internal processes within teams are an important and necessary, but not sufficient, condition to enable coordination in complex multi-team environments.

The role of the Team Leader in coaching for communication both within team and between teams is critically important. This chapter has demonstrated the ways in which effective Team Leaders engaged in coaching for Boundary Riding, Boundary Spanning and Boundary Crossing activities.

When riding the boundaries of the team, leaders are providing feedback and coaching the members to look out for the changing temporal demands associated with the event, and to be mindful of their own responsibilities and how these impact on the responsibilities of others. Facilitating Boundary Spanning to integrate functional unit activities is also important and is undertaken through role orchestration (linking) and blocking so that team members directly engage rather than communication being mediated through the leader. Leaders also coach members to be mindful of the information needs of external stakeholder teams and to manage these expectations and relationships with clarity and assertiveness.

These aspects of communication assist teams to develop collective mindfulness approximate high reliability organising (Weick and Sutcliffe 2001; Barton and Sutcliffe 2009).

Acknowledgements

The research was supported with funding from the Bushfire Cooperative Research Centre. However, the views expressed are those of the author and do not necessarily reflect the views of the Board of the funding agency. I would especially like to thank Kirsty Vogel, Debbie Vogel and Jan Douglas for their help with the data collection and analysis that contributed to the chapter.

References

Barton, M. and Sutcliffe, K.M. (2009). Overcoming dysfunctional momentum. *Human Relations*, 62(9), pp. 1327–1356.

Burke, C.S., Stagl, K.C., Salas, E., Pierce, L. and Kendall, D. (2006). Understanding team adaption: A conceptual analysis and model. *Journal of Applied Psychology*, 91(6), pp. 1189–1207.

Cannon-Bowers, J.A. and Salas, E. (1997). Teamwork competencies: The interaction of team member knowledge, skills and attitudes. In H.F. O'Neil (ed.), *Workforce readiness: Competencies and assessment*. New Jersey: Lawrence Erlbaum Associates, pp. 151–174.

Cannon-Bowers, J.A. and Salas, E. (eds), (1998). *Making decisions under stress: Implications for individual and team training* (1st edition). Washington: American Psychological Association.

Carson, J.B., Tesluk, P.E. and Marrone, J.A. (2007). Shared leadership in teams: An investigation of antecedent conditions and performance. *Academy of Management Journal*, 50(5), p. 1217.

Comfort, L. and Kapucu, N. (2006). Inter-organizational coordination in extreme events: The World Trade Center attacks, September 11 2001. *Natural Hazards*, 39(2), pp. 309–327.

Conger, J.A. and Kanungo, R.N. (1998). *Charismatic leadership in organizations.* Thousand Oaks: Sage.

Crichton, M.T., Lauche, K. and Flin, R. (2005). Incident command skills in the management of an oil industry drilling incident: A case study. *Journal of Contingencies and Crisis Management*, 13(3), pp. 116–128.

DeChurch, L.A. and Zaccaro, S.J. (2010). Perspectives: Teams won't solve this problem. *Human Factors*, 52(2), pp. 329–334.

Edmondson, A.C. (2003). Speaking up in the operating room: How Team Leaders promote learning in interdisciplinary action teams. *Journal of Management Studies*, 40(6), pp. 1419–1452.

Edmondson, A. (2005). Psychological safety and learning behaviour in work teams. In C.L. Cooper and W.H. Starbuck (eds), *Work and workers* (Vol. 3). London: Sage, pp. 291–329.

Edmondson, A.C. and Nembhard, I.M. (2009). Product development and learning in project teams: The challenges are the benefits. *Journal of Product Innovation Management*, 26(2), pp. 123–138.

Fiore S.M., (2008). Interdisciplinarity as teamwork: How the science of teams can inform team Science. *Small Group Research*, 39(3), pp. 251–277.

Flin, R., O'Connor, P. and Crichton, M. (2008). *Safety at the sharp end: A guide to non-technical skills.* Aldershot: Ashgate.

Hackman, J.R. and Wageman, R. (2005). A theory of team coaching. *Academy of Management Review*, 302, pp. 269–287.

Hambley, L.A., O'Neill, T.A. and Kline, T. (2007). Virtual team leadership: The effects of leadership style and communication medium on team interaction styles and outcomes. *Organizational Behavior and Human Decision Processes*, 103(1), pp. 1–20.

Katz, I. and Epstein, S. (1991). Constructive thinking and coping with laboratory-induced stress. *Journal of Personality and Social Psychology*, 61, pp. 789–800.

Langan-Fox, J., Anglim, J. and Wilson, J.R. (2004). Mental models, team mental models, and performance: Process development, and future directions. *Human Factors and Ergonomics in Manufacturing*, 14(4), pp. 331–352.

Lutz, L.D. and Lindell, M.K.(2008). Incident command system as a response model within emergency operation centers during Hurricane Rita. *Journal of Contingencies and Crisis Management*, 16(3), pp. 122–134.

Marcus, L.J., Dorn, B.C. and Henderson, J.M. (2006). Meta-leadership and national emergency preparedness: A model to build government connectivity. *Biosecurity and Bioterrorism*, 4, pp. 128–34.

Mathieu, J., Maynard, M.T., Rapp, T. and Gilson, L. (2008). Team effectiveness 1997–2007: A review of recent advancements and a glimpse into the future. *Journal of Management*, 34(3), pp. 410–476.

Mohammed, S. and Dumville, B.C. (2001). Team mental models in a team knowledge framework: Expanding theory and measurement across disciplinary boundaries. *Journal of Organizational Behaviour*, 22, pp. 89–106.

Moynihan, D.P. (2007). *From forest fires to hurricane Katrina: Case studies of incident command systems*. Washington, DC: Government Office.

Nembhard I.M. and Edmondson, A.C. (2006). Making it safe: The effects of leader inclusiveness and professional status on psychological safety and improvement efforts in health care teams. *Journal of Organizational Behaviour*, 27, pp. 941–966.

Reid, J. (2006). *Addressing lessons from the emergency response to the 7 July 2005 London bombings: What we learned and what we are doing about it*. Report to the British Government. London.

Salas, E., Rosen, M.A., Burke, C.S., Goodwin, G.F. and Fiore, S.M. (2006). The making of a dream team: When expert teams do best. In K.A. Ericsson, N. Charness, P.J. Feltovich and R.R. Hoffman (eds). *The Cambridge handbook of expertise and expert performance*. New York: Cambridge University Press, pp. 439–453.

Salas, E., Rosen, M., Burke, C. and Nicholson, D. (2007). Markers for enhancing team cognition in complex environments: the power of team performance. *Aviation, Space, and Environmental Medicine*, 78(5), pp. 77–85.

Schauboeck, J., Lam, S.S.K. and Cha, S.E. (2007). Embracing transformational leadership: Team values and the impact of leader behaviour on team performance. *Journal of Applied Psychology*, 92(4), pp. 1020–1030.

Scholtens, A. (2008). Controlled collaboration in disaster and crisis management in The Netherlands; History and practice of an overestimated and underestimated concept. *Journal of Contingencies and Crisis Management*, 16(4), pp. 125–134.

Sy, T., Cote, S. and Saavedra, R. (2005). The contagious leaders: Impact of the leader's mood on the mood of group members, group affective tone, and group processes. *Journal of Applied Psychology*, 90(2), pp. 295–305.

Taggar, S. and Seijts, G.H. (2003). Leader and staff role-efficacy as antecedents of collective-efficacy and team performance. *Human Performance*, 16(2), pp. 131–156.

Teague, B., McLeod, R. and Pascoe, S. (2010). *2009 Victorian Bushfires Royal Commission final report*. Melbourne.

Vogus, T.J. and Sutcliffe, K. (2007). The safety organizing scale: Development and validation of a behavioural measure of safety culture in hospital nursing units. *Medical Care*, 45(1), pp. 46–54.

Volpe, C.E., Cannon-Bowers, J.A. and Salas, E. (1996). The impact of cross-training on team functioning: An empirical investigation. *Human Factors*, 38(1), pp. 87–101.

Weick, K.E. and Sutcliffe, K.M. (2001). *Managing the unexpected: Assuring high performance in an age of complexity*. San Francisco: Jossey-Bass.

Wise, C.R. (2006). Organizing for homeland security after Katrina: Is adaptive management what's missing? *Public Administration Review*, 66(3), pp. 302–318.

Yukl, G.A. and Van Fleet, D.D. (1992). Theory and research on leadership in organizations. In M.D. Dunnette and L.M. Hough (eds) *Handbook of industrial and organizational psychology* (Vol. 3). Palo Alto, CA: Consulting Psychologists Press.

Zaccaro, S.J., Ardison, S.D. and Orvis, K.L. (2004). Leadership in virtual teams. In D.V. Day, S.J. Zaccaro and S.M. Halpin (eds), *Leader development for transforming organizations: Growing leaders for tomorrow*. Mahwah, NJ: Psychology Press, pp. 297–262.

Zaccaro, S.J., Heinen, B. and Shuffler, M. (2009). *Team effectiveness in complex organizations: Cross-disciplinary perspectives and approaches*. New York: Routledge/Taylor and Francis Group.

Chapter 8

Firefighter Decision Making at the Local Incident and Regional/State Control Levels

Peter Bremner

Central Queensland University, Appleton Institute, Adelaide, South Australia and Bushfire Cooperative Research Centre, Melbourne, Australia

Dr Chris Bearman

Central Queensland University, Appleton Institute, Adelaide, South Australia and Bushfire Cooperative Research Centre, Melbourne, Australia

Andrew Lawson

South Australian Country Fire Service, Adelaide, South Australia

Introduction

Large-scale emergency fire management is a complex endeavour, with personnel at multiple different levels of an organisation making decisions about how best to fight fires and manage incidents. At the local incident level, firefighters and incident commanders make tactical decisions about fire management. At regional and state level fire officers make strategic decisions to coordinate resources, liaise with external bodies and ensure levels below them are making effective decisions. An effective response to an emergency, therefore, depends on people at all levels of fire management making good decisions. Understanding how people make decisions at regional and state levels of emergency management is particularly important given that post-accident and coronial inquiries often focus on decisions made at these levels following large-scale emergencies.

This chapter considers research on the kinds of decisions that are made at local and regional/state levels of fire management, discusses how people make decisions at these levels and examines some of the pressures that may lead to impaired decision making. While there is some research into these topics at local incident management levels (e.g., McLennan and Omodei 1996; Bearman and Bremner 2013), there is little information about decision making at regional and state levels. We will therefore discuss the literature on firefighter decision making at local levels and consider how this might apply to regional/state levels of emergency fire management. In addition we will highlight areas where more research needs to be undertaken. To provide a background for our consideration of local and regional/state level decision making we first briefly discuss the management of the emergency response.

Emergency Response

Within Australia, the Australasian Inter-Service Incident Management System (AIIMS) developed by the Australasian Fire and Emergency Service Authorities Council (AFAC) provides a common management framework for incident management. AIIMS contains processes and procedures that can be applied to incidents of all sizes and it 'provides the basis for an expanded response as an incident grows in size and complexity' (AFAC 2013, p. 2). Within AIIMS, the key roles for emergency incident management vary from small incidents with one Incident Controller to very large incidents where the fireground is separated into divisions and sections and key incident management functions (planning, intelligence, public information, operations, investigation, logistics and finance) are delegated to other people. For all levels of incident, the Incident Controller has the overall responsibility for controlling and managing the incident.

The AIIMS structure classifies levels of incidents as Level 1 (the Incident Controller controls a small incident using local resources), Level 2 (the Incident Controller manages multiple crews or sectors) or Level 3 (multiple incidents involving many crews, spread across large areas, many resources and multiple external agencies) (AFAC 2013). A fully expanded structure is illustrated in Figure 8.1.

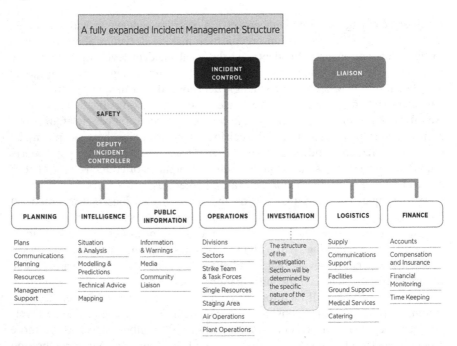

Figure 8.1   The fully expanded Incident Management Structure reproduced with permission from AFAC

For a small emergency, the single Incident Controller will undertake all of the emergency management functions (planning, intelligence, public information, operations, investigation, logistics and finance). As the incident grows in size or complexity some of these functions may be delegated to other individuals or crews. The organisation will support the Incident Controller by providing additional resources that are requested (e.g., additional crews, water, food, aircraft or specialist crews). The Incident Controller may also request that the local incident management team (or higher level coordinator if there is no local incident management team set up) assist by contacting other agencies, including: utility companies, police, and ambulance. The Incident Controller operating at the local incident management team level will make all the decisions required to assess and manage the incident (AFAC 2013). The decision making process for the Incident Controller can be extensive and will involve many decisions from the time of the alarm to conclusion of the incident (Bearman and Bremner 2013).

If the incident escalates, additional resources (such as more crews, water supplies for firefighting and water bombers) will be brought in to assist firefighting operations and a local incident management team may be set up. If the incident continues to escalate it may develop into a Level 2 incident, and the incident management structure will be expanded by the delegation of functions such as logistics and planning to meet the levels of management required to manage the larger size, resources and risks involved. These changes to incident control are made through consultation with the chain of command.

For a Level 3 incident, regional and state level Incident Controllers support multiple incident management teams. At the incident level local Incident Controllers make tactical decisions about fire suppression and control in conjunction with personnel dealing directly with the incident. More strategic decisions about overall resource allocation and large-scale planning are made at regional and state levels. The response to Level 3 incidents is likely to include multiple different agencies, and liaison with government, the media and the public, which is managed at regional and state levels. The chain of information flow for a Level 3 incident is shown in Figure 8.2.

As an emergency escalates there are organisational processes that enable regional and state level Incident Controllers to be introduced into the incident management structure. Certain triggers indicate that regional and state coordination centres should be opened and incident management teams should be formed. Such triggers include: alerts by phone calls; radio communications and pager messages; and the release of the fire danger information for the current or following day.[1] The release of fire danger information and the declaration of a fire ban day will

---

1 Fire danger information includes the fire danger index (FDI), which is an index between 0 and 100 indicating how a bushfire will react based on forecast environmental conditions including temperature, relative humidity and wind speed. The fire danger rating (FDR) is released to the public to communicate the level of bushfire risk and it will have ratings from low-moderate (where fires can be easily controlled) up to the highest

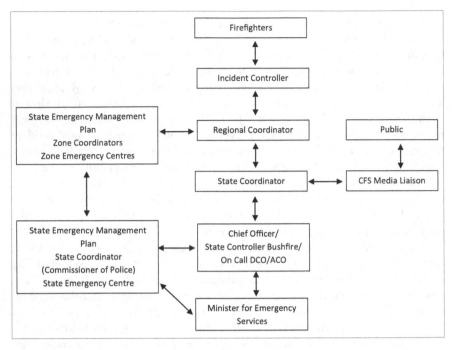

Figure 8.2   Chain of information flow for Country Fire Service operations (Chief Officer's standing orders: Standard operating procedures and operations management guidelines, 2011)

provide triggers to regional staff to enable them to provide the required 'level of preparedness' by setting up and deciding on the level of activation of regional incident management centres in advance. This level of activation may be full or partial (where only a small regional incident management team is needed).

During an emergency response, clear and effective decision making is required at each level of fire management. When decision making is ineffective at any level of fire management an impaired operational response can result. It is important then to understand the kinds of decisions that are made at these different levels, how people make decisions and the pressures that may lead to impaired decision making. We begin by reviewing research on the roles of local incident and regional/ state level fire controllers and the different types of decision that are made at these different levels.

(catastrophic, where fires will be potentially uncontrollable). Based on this information a day of total fire bans (also called a Fire Ban Day) may be declared.

The Roles of Local Incident and Regional/State Level Incident Controllers

This chapter focuses on decision making at two main levels of emergency management: the local incident level and the regional/state level. The local incident level includes the Incident Controller and people below the Incident Controller who are managing the emergency. The regional and state controller roles will be discussed together here because the roles are broadly similar in most Australian states, with the main difference being one of level (i.e., region or state). It is also the case that some Australian states don't have a regional level in their emergency management structure.

Bearman and Bremner (2013) conducted a task analysis to identify the tasks that a volunteer Incident Controller needs to carry out during a Level 1 fire incident. The task analysis identified nine high-level tasks:

1. go to fire station
2. prepare for fire station departure
3. travel to incident
4. arrive at incident
5. implement plan of attack
6. prepare to return to station
7. return to station
8. close station
9. return home.

Within these high-level tasks a number of sub-tasks need to be carried out. For example, for the high-level task of 'travel to incident' the sub-tasks are:

1. check crew are ready
2. check driver is ready and the incident location is known
3. determine incident priority level
4. depart station
5. advise base of departure
6. assist and monitor driver en-route
7. brief crew and base en-route
8. review crew roles and tasks en-route.

Each of these tasks requires the Incident Controller to make decisions. In total, Bearman and Bremner (2013) identified 67 tasks that the Incident Controller needs to conduct. Of these 67 tasks, nine are considered to be particularly important since poor decisions made during these tasks can lead to errors with catastrophic outcomes. These nine tasks highlight decision making that relates to crew management, risk management, tactical fire management and responsibility for crew safety.

Regional and state level personnel are nearly always located away from the emergency and the emphasis at these levels is primarily on strategic fire management. A task analysis for a regional Incident Controller managing a Level 3 incident is shown in Figure 8.3.

The first task is to decide on the size and the composition of the initial regional incident management team based on factors that include the size of the incident and the prevailing weather conditions. As the regional incident management team starts up, the regional Incident Controller will actively seek more information to maintain the team's situational awareness and will review the structure to ensure the team has the resources to fulfil their responsibilities in managing the incident. During the incident, the regional Incident Controller will decide when fire warning messages are to be issued to the public and the type of fire warning messages

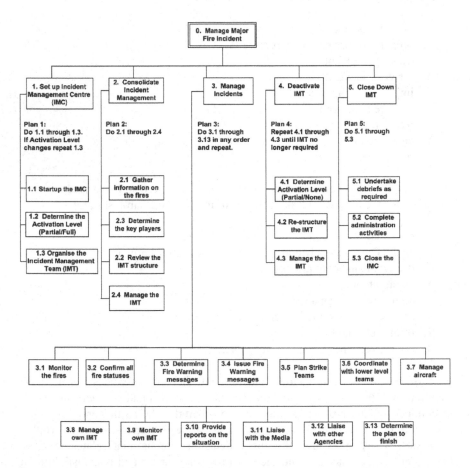

Figure 8.3   Task analysis for the regional Incident Controller in a Level 3 incident

to be sent. These messages will provide advice of the current fire situation and the actions the public should take if they are in the vicinity of the fire. The regional Incident Controller will monitor the effectiveness of the regional incident management team and may make decisions to change the structure or to change personnel within the team. As the incident reduces in size or complexity, the regional Incident Controller will decide if the team may be reduced in size or if it is needed at all.

Having examined the kinds of decisions that are made by local and regional/state level controllers we now turn to how people make decisions in these roles.

Firefighter Decision Making

Gary Klein (2003) has conducted extensive research into the decision making processes of urban firefighters. Klein's research examined the decision making of experienced firefighters performing realistic tasks in a naturalistic decision making setting. Klein (2003) suggested that firefighters make decisions 'largely through a process based on intuition' (p. 21). The process of intuitive decision making is the detection of cues that enable us to match aspects of a situation to patterns that we have experienced in the past.

Based on this research, Klein developed the recognition-primed decision (RPD) model. This model describes how 'decision makers size up the situation to recognise which course of action makes sense, and the way they evaluate that course of action by imagining it' (Klein 1998, p. 24). Klein describes the model as 'a two-part process of pattern matching and mental simulation' (p. 27) and noted that his 'data analysis showed that fire-fighters used the RPD process for more than 80 per cent of their toughest incidents' (p. 28).

Klein's (1998) recognition-primed decision model provides three variations, which are based around the firefighter's recognition of the situation. The first variation occurs when a situation is experienced by the firefighter and is recognised as familiar. Recognising the situation generates expectancies about:

1. that situation
2. relevant cues that should occur
3. plausible goals that might be pursued
4. typical actions that have been carried out in the past.

Based on these cognitive processes, a course of action is implemented.

The second variation occurs when a situation is not recognised as familiar. When this occurs the firefighter will need to go through a process of story building to ascertain whether there are any familiar components to the situation that can be used to support decision making. The third variation occurs when no plausible actions that have been carried out in the past are generated. When this occurs the firefighter needs to mentally simulate solutions that are generated to determine if they are going to work.

In addition to being used at local incident levels, there is some evidence that RPD is also used at regional/state levels of fire management to identify cues that might indicate errors, violations and breakdowns. This research is discussed further in the section on situation awareness.

Martin, Flin and Skriver (1997) suggested that recognition-primed decision making is one of several decision strategies that could be used in an emergency situation. Martin et al. (1997) proposed that there is a continuum of decision strategies with intuitive decision strategies (such as recognition-primed decision) at one end and analytical strategies (such as choice-based decision making) at the other. For emergency management the nature of the situation has a strong influence on which type of strategy is adopted. If a situation is high-risk and time-critical then recognition-primed decision making is more likely to be adopted. If a situation is of lower risk and more time is available then analytic strategies are more likely to be employed. Decision strategies may also change as an incident progresses so that at the beginning of an incident intuitive strategies are more likely to be used and later, as an incident develops, analytic decision strategies are more likely to be used. Fredholm (1997) noted that fire commanders may not always use appropriate decision strategies given the features of the situation, with some firefighters using intuitive strategies when analytic strategies would be more appropriate.

The research of Martin et al. (1997) and Fredholm (1997) raises interesting questions for regional/state levels of decision making. It would be expected that Incident Controllers at regional/state levels would be more likely to use analytic decision strategies rather than intuitive strategies since the situation features will generally be more compatible with analytic strategies. As discussed in the section on situation awareness (below) there is some evidence that regional Incident Controllers use strategies that are consistent with recognition-primed decision making, although the frequency of such decisions compared to analytic strategies cannot be ascertained. Further research is required to investigate the extent to which regional/state Incident Controllers use intuitive and analytic decision strategies compared to the features of the situation.

McLennan and Omodei (1996) suggested that an important process that underpins emergency management decision making is pre-priming. Experienced decision makers can anticipate the emergency situation based on prior experience and prime a number of most-likely prototypical situations. McLennan and Omodei argued that situational assessments and possibilities for action are first made from among this reduced set of pre-primed prototypes. For personnel operating at regional/state fire management levels a similar process is likely to occur based on initial information about the emergency. However, further research is required to determine the extent to which pre-priming occurs and the way it is used at this level of emergency management.

Situation awareness

An important part of decision making (and many models of decision making, including RPD) is building and maintaining a good understanding of the ongoing situation, also known as maintaining situation awareness. Without good situation awareness it is difficult to make effective decisions because the information on which those decisions are based may be incorrect or not current.

Endsley (1988) defined situation awareness as, 'the perception of the elements in the environment within a volume of time and space, the comprehension of their meaning and the projection of their status in the near future' (Endsley 1988, p. 97 in Endsley and Garland 2000, p. 5). Thus, Endsley identifies three levels of situation awareness: perception, comprehension and projection. The first level, *perception*, is described as 'perceiving the status, attributes, and dynamics of relevant elements in the environment' (Endsley et al. 1998, p. 1). For an Incident Controller this would include information about the response (number of trucks, crew members, etc.) and information about the external environment (weather, fire behaviour). Endsley described the second level of situation awareness as *comprehension*, which enables interpretation of the meaning and significance of the information. Endsley et al. (1998) noted that this also includes forming a picture of the environment, and more experienced people may comprehend the situation better than novices. The third level of situation awareness, *projection*, is described by Endsley as future projections or forecasting future events. These projections are important in determining fire behaviour based on weather changes and identifying possible issues that may develop.

Examples of each of Endsley's categories of situation awareness could be found in data that we have collected on the challenging situations faced by volunteer Incident Controllers. As part of a broad study of the tasks, pressures and mitigating strategies of volunteer Incident Controllers we have conducted 10 critical decision method interviews with volunteer Incident Controllers in Australia. In a critical decision method interview participants are asked to describe situations that were particularly challenging for them (Crandall, Klein and Hoffman 2006). These incidents are then probed by the researcher using non-directional probes (e.g., what was happening at the time, what were the key decisions made and why) to elaborate details about the incident. The critical decision method interviews that we collected with volunteer Incident Controllers formed part of the data for the task analysis conducted by Bearman and Bremner (2013) discussed earlier and more details on the method can be found in their paper. As part of the data collection we asked questions about how the participants built and maintained an understanding of the situation during these challenging events. To provide examples of Endsley's categories of situation awareness for this chapter we conducted a preliminary top-down analysis of the transcribed data by placing extracts into the three situation awareness categories: perception, comprehension and prediction. Examples of each of the categories are presented below.

At the perception level, Incident Controllers discussed elements of the current situation:

> Get the tools out of the toolbox and say, OK what's the slope, what's the
> vegetation, fuel loading is it elevated fuel, ground fuel, continuous, you know
> has it got vertical or horizontal continuity, what's the weather, I mean is it really
> quite dry.

At the comprehension level, Incident Controllers discussed their understanding of
the situation:

> It's those little variables like the wind direction, what's going on there, whether
> the fuel is burning at a higher intensity than what I think it's it should have burnt
> at, yeah how high the flame height is.

At the prediction level, Incident Controllers discussed their prediction of future
states of the fire.

> Where are we now, where's it going and what's it going to impact because that's
> how I determine how I go about fighting a fire.

To maintain appropriate situation awareness, the local Incident Controller on the
fireground needs to monitor the crew, the fire and the surrounding environment.
During these tasks the Incident Commander must perceive, comprehend and make
predictions based on local information in their area of operations. Such local
information includes: the local weather conditions and forecast; the geography
of the area; the density and type of the bush; the location of crew members, the
locations of local assets (such as houses, farms and buildings) and the locations
of nearby resources (such as water supplies). The Incident Controller at or near
the fire needs to manage all of the gathered incident information and actively
seek any key information that may be missing from their situational picture. This
information set is necessary to maintain situational awareness and to provide
effective situation reports (see also Owen, Chapter 7, this volume for further
discussion on the importance of effective communication and high-reliability).

Regional and state level Incident Controllers are remote from the fireground
and are unable to use the same cues that are used by local Incident Controllers to
develop situation awareness. Instead regional or state level Incident Controllers
need to develop situation awareness through situation reports, communication
with people who are closer to the fire and through their own experience of the way
fires generally develop. Given the role of regional/state Incident Controllers and
the large amount of information that is available from different sources it is likely
that an important part of maintaining situation awareness at regional and state
levels is synthesising information and detecting anomalies.

In an unpublished study by Grunwald and Bearman (2013) regional Incident
Controllers reported several different types of methods to detect cues that

may indicate an error, violation or breakdown[2] has occurred at lower levels of emergency management. These cue detection methods are:

1. inconsistent information-based
2. intuition-based
3. network-based
4. proxy-based.

The four cue detection methods help the regional Incident Controllers form their ongoing situation awareness and potentially indicate that something is amiss.

Regional Incident Controllers use information-based cue detection when information from situation reports, data analysis, or factual information from communication contains discrepancies or is inconsistent with what the regional Incident Controller thinks should be occurring. When information is inconsistent a problem with the local level emergency management team might be indicated.

Regional Incident Controllers are also able to detect cues through intuition. Regional Incident Controllers are generally very experienced and can use this experience to 'sense' that there is a problem in the way the team is operating, even when there are no explicit cues they could point to. According to Klein's (1998) recognition-primed decision model (discussed earlier), such intuition occurs when a pattern in a situation is recognised as familiar but the particular situation that was experienced before is not retrieved from memory. The recognition triggers expectancies and cues, which if they do not match current experience, indicates to the person that something is wrong, without providing an explicit reason why.

Regional Incident Controllers also detect cues through information passed through their informal and formal networks, and through the use of proxies, that is, designated people who act on behalf of the regional Incident Controller.

The discussion so far has presented a somewhat idealised notion of how people make decisions. However, the firefighting environment contains numerous pressures that may influence the way that fire personnel at different levels make decisions. The next section discusses some of these pressures together with examples from the domain of emergency management.

Pressures on Decision Making

It is likely that people at all levels of fire management experience pressure from the external environment, the organisation, the social environment, the ongoing situation, and individual physical pressures (see McLennan et al., Chapter 2, this

---

2   Theoretical distinctions have been drawn between errors and violations based on the intention of the person (Reason 1990) and between errors and breakdowns based on individual versus system processes. However, such distinctions need not concern us here since errors, violations and breakdowns were detected in a similar way in this study.

volume; Douglas, Chapter 5, this volume; Brooks, Chapter 9, this volume). We will provide examples of pressures based on local incident management levels and discuss what is likely to occur at regional/state levels.

A Framework for Understanding Pressures

One way of conceptualising the pressures that can be exerted on a person in an operational environment is through a framework like the Human Factors Analysis and Classification System (HFACS) (Wiegmann and Shappell 2003; Shappell and Wiegmann 2005; Patterson and Shappell 2010). HFACS was developed to determine the causes of aviation accidents, and has been used to examine large-scale bushfires. Despite the focus on accident analysis, HFACS represents a fairly simple practical framework that has identified the key influences that can cause a person working in a high-reliability organisation to make an error or commit a violation. As such it is a useful framework for thinking about the potential causes of decision errors and violations in emergency management. Wiegmann and Shappell (2003) described the factors in the framework as latent vulnerabilities; however, following Paletz et al. (2009) we prefer to use the more neutral term *pressure*. In contrast to latent vulnerabilities, pressures may be 'strong or weak, subtle or coercive, direct or indirect' (Paletz et al., p. 436). Pressures are also not necessarily bad, some pressures merely exist in a person's operational world.

The Human Factors Analysis and Classification System describes different levels of influence on a person. The original version of HFACS is a linear progression from organisational influences, to unsafe supervision, to preconditions for unsafe acts, to the unsafe acts themselves. Two of the authors (Bearman and Bremner) have redesigned HFACS as a circular model (CHFACS), which emphasises that each of the factors may have a direct influence on a person and may interact with each other (see Figure 8.4). Bearman and Bremner were particularly interested in decision making so they placed decision making at the centre of the model. A person's decision making may be influenced by a number of types of pressures, which include:

1. environmental factors
2. conditions of operators
3. personnel factors
4. strong situations
5. supervision issues
6. organisational issues.

Environmental factors, conditions of operators and personnel factors would be described as preconditions for unsafe acts by Wiegmann and Shappell (2003).

To explicate some of the pressures that might lead to faulty decisions we examined the data from the 10 critical decision method interviews with volunteer

Incident Controllers discussed earlier. In addition to containing information about the tasks required by the volunteer Incident Controller and information about situation awareness, the critical decision method data also contained information about the pressures faced by firefighters. Two of the authors (Bearman and Bremner 2013) have explored one particular type of pressure in this data, the pressures from strong situations (i.e., goal seduction and situation aversion); however, a number of other pressures could also be identified.

To explore the pressures in the data a bottom-up and top-down analysis process was used. First, the transcribed interviews were analysed by cycling through the processes of documentation, extraction, classification and synthesis. This process produced information in a bottom-up, data-driven manner about the pressures faced by Incident Controllers and crews. These pressures were then coded into the CHFACS categories, which is a top-down concept driven data analysis. The data analysis focused on pressures faced by both the volunteer Incident Controllers and the crew. Pressures on the crew were included since managing pressures faced by crew members is an important role of the Incident Controller. The CHFACS categories that could be identified in the data are discussed below. It should be noted that this preliminary analysis is not meant to be an extensive or exhaustive list. The data presented here is designed to exemplify the categories and to highlight some of the pressures that are faced by volunteer Incident Controllers and their crew. Further research is required to explore these pressures in more detail and to examine how frequently they occur.

Figure 8.4   The Circular Human Factors Analysis and Classification System (CHFACS)

Environmental factors

Three types of environmental factors have been identified: physical environment, technological environment and social environment (Wiegmann and Shappell 2003; Paletz et al. 2009). Physical environment refers to the operational environment (such as weather and terrain) and the ambient environment (such as heat, light and smoke). The physical environment in particular often exerts pressure on volunteer firefighters given that a large part of their role is to fight fires in the bush. On occasions these conditions can be particularly tough and this can exert considerable pressure on the firefighter.

> ...and that was particularly challenging...it was it was very very windy...and the terrain, a lot of the terrain was very difficult to access...I think we were sort of on the fireground that day for about 16 hours without much of a break.

> ...especially in a hot situation because the last thing you want, and we've had it happen, is people going down from heatstroke, and then that creates another problem.

> I think one nightshift is one of the toughest of the lot as far as I'm concerned...your reflexes and everything else have got to be...50 per cent better than dayshift.

The technological environment refers to the equipment required to manage fires. While it is likely that there will be pressures from the technological environment, (such as the design of equipment) there were no examples of this in the data.

The social environment concerns the direct and indirect pressures that can be exerted by people outside of the fire management team (Paletz et al. 2009). There were several comments in the data about the pressure that can be placed on firefighters by incident participants and members of the public.

> A vehicle accident you may have somebody seriously injured or trapped and other occupants of the car are not seriously injured or trapped but they are traumatised by the fact that one of their relatives or friends are in a precarious situation and they can put a lot of pressure on particularly the crew leader that you're not doing enough quickly enough.

> I've had someone screaming in my ear – the horses and the sheep they're over the road the fires going to be in my paddock in a minute and don't worry about this job, you come over here, and so that causes a bit of pressure.

> The pressures are on me were actually external by the neighbours from each side of that particular property knowing that he was in there and wanting, er, someone to rescue him, umm, and you know when those people are sort of yelling out 'he's in there, he's in there'.

> Crew are relatively easy to deal with in that situation, general public are very
> hard and you can exhaust a lot of energy and get no result.

Since regional/state level fire controllers are nearly always located away from the emergency in an office-like environment these pressures are unlikely to influence regional/state level decision making.

Condition of the operator

The condition of the operator can have a large influence on their decision making (Wiegmann and Shappell 2003). Three types of operator conditions have been identified that can lead to poor decisions and violations: mental states, physiological states and mental/physical limitations.

Mental states are pressures caused by the person being in a particular mental state or condition. In contrast, physiological states are pressures caused by the physiological condition of the person. In the examples in the data that we are considering here it was not always clear whether the pressure was specifically from the mental or physical state, so we will discuss these categories together.

> Because they've done all the hard work, they're sort of semi worn out and all
> of a sudden that's when accidents happen you know they'll go and walk into a
> cupboard door or getting up the step they'll miss the step.

> Just being aware that everybody is knackered because when you do come off
> an adrenalin high you come down pretty hard and that everybody else is just
> basically going to sleep.

> So, yeah, the physical pressures are pretty demanding. You need to be able to
> monitor how your body feels because nobody else knows.

There were no examples that referred to mental or physical limitations in the data.

Personnel factors

Personnel factors refer to pressures that are due to actions carried out by people. Two types of personnel factors have been identified: individual fitness for duty and team coordination (referred to as crew resource management in Wiegmann and Shappell 2003 and communication and coordination in Patterson and Shappell 2010).

In a fire management situation team coordination is crucially important and was identified by Bearman and Bremner (2013) as a set of tasks that could have very serious consequences if errors were made. Crew communication in particular was highlighted as a task with a high risk rating since poor communication might lead to 'reduced crew situational awareness, which can increase the likelihood of confusion, sub-optimal crew decision making, and add delays in implementing actions' (Bearman and Bremner 2013, p. 491). In relation to crew communication, one of the participants in Bearman and Bremner (2013) stated 'the worst thing is not letting your crews know' (p. 492).

Another pressure for the Incident Controller is to ensure that the crew composition is correct for the incident.

> There is a bit of pressure on the…officer in charge to make sure that you know he has got enough crew, he's got crew, the right crew for the situation.

One of the important crew role decisions was to allocate the driver of the vehicle. This decision was based on factors including experience and safety.

> It is primarily the first choice for me is always who is my most experienced driver because we have to get there.

It should be noted that allocation of crew roles is often considered to be a task for supervisors in Wiegmann and Shappell's (2003) coding scheme. In contrast to aviation, where Wiegmann and Shappell developed the Human Factors Analysis and Classification System, emergency management has more people involved and a more complex structure. Since the focus in this data was on the Incident Controller crew allocation is included in this section, with supervision issues reserved for the interaction of the Incident Controller with regional and state levels of emergency fire management.

It seems likely that regional/state level fire controllers will also face pressure from team coordination and that this will manifest itself slightly differently because of differences in the environment in which these teams operate. Further research is required to investigate this.

There were no examples of firefighters not being fit for duty.

Strong situations
Strong situations are situations that exert too much influence on decision making (Bearman, Paletz and Orasanu 2009). Bearman and colleagues (Bearman et al. 2009; Bearman and Bremner 2013) have identified two types of strong situations: goal seduction and situation aversion. Goal seduction is described as an attraction towards a goal, while situation aversion is described as repulsion away from a situation. Bearman and Bremner (2013) have identified three types of goal seduction situations in firefighting: getting to the fire station quickly, getting to the incident quickly, and wanting to begin dealing with the incident straight away. Each of these types of goal seduction, also discussed in Bearman and Bremner (2013), is outlined in the quotes below:

> Because they are only thinking of one thing and the adrenalin is rushing around and you just want to get to the station as quick as possible and I must admit I'm guilty of that sometimes.

> The biggest drama people have is that you know it's an emergency an emergency
> we got to rush to it we got to rush to it.

> Structure fires are a classic example and you get blokes that, er, will just turn
> up or appliances just turn up and they'll barge off of the truck, grab a hose and
> disappear into the smoke.

Situation aversion was also identified, particularly in relation to the end of
an incident:

> For me it's to make sure everything is fully extinguished or fully complete as
> far as the situation goes because everybody wants to go home, no-one wants to
> do the hard yards.

It is unlikely that regional/state levels of fire management will be subject to similar
manifestations of goal seduction and situation aversion because of the different
environment in which they work. It is not clear if and how these concepts would
occur in a coordination role that does not require physical actions to fight the fire.

Supervision issues

Pressures on an incident commanders' decision making can come from
regional and state levels of the organisation. Wiegmann and Shappell (2003)
identified four categories of supervision issues: inadequate supervision, planned
inappropriate operations, failure to correct a known problem, and supervisors
violating procedures.

The category of planned inappropriate operations is concerned with operational
decisions at higher levels of the organisation that are inappropriate in some
way. Examples in the data were based on poor decisions (from the participants'
perspective) by more senior officers with a supervisory role who may have limited
situation awareness:

> Yeah we did a lot of work and they wouldn't let us come home...until we had a
> sleep. You know how the hell do you sleep in a tent when it's 40° at (location)?

> If you're out on the fireground and you've made a decision on something but if
> you get told from up high you need to do this and...you know that's not the case
> and you then have to justify to these other people who outrank you.

> There's always a great deal of pressure brought to bear by communication, that
> you are dealing with people on a radio who aren't at the incident.

There were no examples in the data of inadequate supervision, failure to correct a
known problem, or supervisory violations.

These examples show the interaction between the regional/state and local incident management levels from the perspective of local incident management and show some of the challenges that are faced in coordination between local incident and regional/state levels. While these quotes may indicate the difficulties of maintaining situation awareness at regional/state levels of fire management this should be interpreted cautiously since we don't have the regional/state controller's perspective. It is likely that regional levels of fire management would experience supervision issues with state levels, although its not clear what form these would take without further research.

Organisational influences
Organisational influences relate to decisions made by the upper levels of an organisation. Three different types of organisational influences have been identified: resource management, organisational climate, and organisational processes. In an emergency management context, resource management encompasses the regional/state level decisions regarding the allocation of human and physical resources. Examples of the pressures that resource allocation exerted on decision making by incident commanders could be found in the data:

> So as the officer in charge not only am I trying to coordinate the rescue operation, the firefighting operation and also the request for additional resources, I'm now concerning myself from the fact that we were being supported by a...brigade that ... didn't have the trained people or the equipment to actually provide the essential support that we needed.

The organisational climate will dictate the attitudes of the fire management personnel to safety. A number of quotes were found in the data that suggested that at times attitudes to safety were not always what they could be:

> There's sometimes that I've seen people with my own brigade and in other brigades their focus sometimes isn't always on safety, their focus is on perhaps suppression of the fire at all costs sometimes.

> And that's where I've had clashes through knowledge and through safety principles.

> I gave orders to the crews so that it is an electrical fault stay right away from it until (power company) arrived. But, umm, good old [firefighter name] he just took the crews up to the house window and, umm, around the corner to have a look anyway.

Regional/state levels of emergency management make decisions about resource allocation and organisational processes and the pressures experienced by local Incident Controllers are often because of these decisions. This again shows the interaction between the different levels of emergency management. It is likely that regional/state level Incident Controllers will face pressures from organisational factors, particularly organisational culture. However, the way in which these pressures manifest themselves and the influence that they have is currently unknown.

Outside Factors

Outside factors are factors from outside the organisation (such as regulations, economic pressures, environmental concerns and legal pressures) that may have an effect on an organisation's operations (Patterson and Shappell 2010). While there was no data in the present analysis that fell into this category it is likely that outside factors exert considerable pressure on state and regional levels of fire management given the interaction of these levels with the media, public and political representatives. More research needs to be conducted into how these pressures influence this level of fire management.

Implications of Pressures for Practitioners and Instructors

This preliminary analysis has highlighted some of the many pressures that may influence decisions made by fire personnel. It is likely that pressures will occur in combination so that a fire Incident Controller must deal with several pressures at a time. For example, a fire Incident Controller may be fatigued, and feel a strong desire to start managing the incident and be subject to pressure from people involved in the emergency. This kind of situation represents the operational reality of local incident management and represents the context in which decisions are often made. While we have emphasised that such pressures can lead to poor decisions and violations in this section it should be noted that in the large majority of cases firefighters deal with these pressures successfully and usually provide a highly effective response to the emergency. While to some extent pressures are part of the operational reality of managing emergencies that shouldn't prevent us from seeking to reduce the effects of pressure where possible. The more we understand about the effects of pressures the easier it is to develop procedures and cognitive strategies for minimising their effects so that poor decisions and an impaired operational response are less likely to occur.

Conclusions

Large-scale fire management is a complex endeavour with personnel at multiple different levels of an organisation making decisions about how best to manage that emergency. An effective response to an emergency depends on people at all levels of fire management making effective decisions. This chapter has outlined the different roles that people have in different types of emergencies; examined the roles and key decisions of people in those roles; discussed how people make decisions and maintain situation awareness; and identified some of the pressures that can lead people to make poor decisions. It is clear from this chapter that we currently know very little about decision making at regional/state levels of fire management. It is unclear what decisions are made, how they are made or what pressures may lead to faulty decision making. Given the focus of post-accident and coronial inquiries on decisions made at regional/state levels of emergency management following large-scale emergencies it is vitally important that we rectify this knowledge gap to better support regional/state level decision making.

Acknowledgements

The research was supported by the Bushfire Cooperative Research Centre Extension Grant. However, the views expressed are those of the authors and do not necessarily reflect the views of the Board of the funding agency.

References

Australasian Fire and Emergency Service Authorities Council (2013). *The Australasian Inter-service Incident Management System (AIIMS)* 4th edition 2013 Revision.

Bearman, C.R. and Bremner, P.A. (2013). A day in the life of a volunteer incident commander: Errors, pressures and mitigating strategies. *Applied Ergonomics,* 44, pp. 488–495.

Bearman, C.R., Paletz, S.B. and Orasanu, J. (2009). Situational pressures on decision making: Goal-seduction and situation aversion. *Aviation, Space & Environmental Medicine*, 80, pp. 556–560.

Crandall, B., Klein, G. and Hoffman, R.R. (2006). *Working minds: A practitioner's guide to cognitive task analysis.* Cambridge, MA: The MIT Press.

Endsley, M.R. (2000). Theoretical underpinnings of situation awareness: A critical review. In Endsley, M.R. and Garland, D.J. (eds), *Situation awareness analysis and measurement.* Mahwah, NJ: Lawrence Erlbaum Associates Inc., pp. 3–32.

Endsley, M.R., Farley, T.C. et al. (1998). *Situation awareness information requirements for commercial airline pilots.* International Center for Air Transportation.

Fredholm (1997). Decision making patterns in major fire-fighting and rescue operations. In Flin, R.H., Salas, E., Strub, M. and Martin, L. (eds). *Decision making under stress: Emerging themes and applications.* Aldershot: Ashgate.

Johnson, C., Cumming, G. and Omodei, M. (2008, September). How worst case scenarios are considered by bushfire fighters: An interview study. In *Proceedings of the International Bushfire Research Conference.*

Klein, G. (1998). *Sources of power: How people make decisions*. Cambridge, MA: MIT Press.

Klein, G. (2003). *The power of intuition*. New York: Doubleday.

Martin, L., Flin, R. and Skriver, J. (1997). Emergency decision making: A wider decision framework? In Flin, R.H., Salas, E., Strub, M. and Martin, L. (eds). *Decision making under stress: Emerging themes and applications*. Aldershot: Ashgate.

McLennan, J. and Omodei, M. (1996). The role of prepriming in recognition-primed decision making. *Perceptual and Motor Skills*, 82, pp. 1059–1069.

Paletz, S.B.F., Bearman, C.R. et al. (2009). Socialising the human factors analysis and classification system: Incorporating social psychological phenomena into a human factors error classification system. *Human Factors*, 51(4), pp. 435–445.

Patterson, J.M. and Shappell, S.A. (2010). Operator error and system deficiencies: Analysis of 508 mining incidents and accidents from Queensland, Australia using HFACS. *Accident Analysis & Prevention*, 42(4), pp. 1379–1385.

Shappell, S.A. and Wiegmann, D.A. (2005). *Human error and general aviation accidents: A comprehensive, fine-grained analysis using HFACS*. Office of Aerospace Medicine: Washington, DC.

Wickens, C., Lee, J. et al. (2004). *An introduction to human factors engineering*. 2nd edition. Upper Saddle River, New Jersey: Pearson.

Wiegmann, D.A. and Shappell, S.A. (2003). *A human error approach to aviation accident analysis*. Aldershot: Ashgate.

Chapter 9

Coping Ugly: Errors, Decisions, Coping and the Implications for Emergency Management Training

Dr Benjamin Brooks

Australian Maritime College, University of Tasmania, Australia
and Bushfire Cooperative Research Centre, Melbourne, Australia

Introduction

Training and exercising are key components of a strategic move forward in effective organising above the Incident Management Team level. These activities support managers and coordinators to build and maintain a range of skills that are both technical (e.g., fire/flood behaviour) and non-technical (leadership, communication) in order to provide the most effective emergency management coordination possible. To this point there has been far greater emphasis on training at the Incident Management Team level and below than there has been above.

The aim of this chapter is to identify if there are opportunities to enhance training efforts for those who inhabit these roles. To achieve this, it is necessary to explore the nature of non-technical skills and competencies involved with coordination of emergency management events; identify how these skills/competencies might break down; and establish the range of formal and informal training activities that currently occur.

It is also necessary to acknowledge that emergency management training and exercising activities face some significant constraints and these in turn constrain what can reasonably be proposed as a training pathway. Jurisdictional approaches above Incident Management Teams are not uniform, opportunities and resources for training above the Incident Management Team level are not infinite. Emergency events are changing in terms of complexity, intensity and duration. In addition, public and political expectations of managers are challenging, if not sometimes impossible to meet.

In Australasia current training systems application of human factors related skills and knowledge are neither universally poor, nor systematically complete. Murphy and Dunn (2012) recently collated the Noetic Group's experience with lessons learnt studies and post-activity reviews for a range of emergency management events. They concluded that a pattern of leadership failure emerged from this analysis:

The failure is seldom one of character, but inevitably a lack of preparation and understanding. Leaders, and their teams, are unable to effectively apply their knowledge and skills to a situation that is either so novel, or of a scale that is beyond their experience and conception. (p. 2)

However, the same authors also made the following comments regarding training issues:

After examining several disasters, it is clear there has been a lack of resources and insufficient attention given to training. The Noetic Group found that response training for routine accidents is effective at all levels. However, this is not the case for novel or 'out of scale' disasters. (p. 7)

These are not insignificant challenges. The complexity of a State Control Centre includes multiple agencies and hundreds of individuals, making thousands of decisions in sometimes information-rich, and other times information-poor environments. Thankfully, emergency management agencies are not the first work domain to proceed down this path. High risk transportation domains such as aviation, maritime and rail (both passenger and bulk) have addressed similar problems, and therefore at least provide a model of how emergency management agencies might proceed.

A typical starting point has been to establish how the system breaks down, and understanding in this area has often been derived from detailed accident investigations or other assessments of human error.

Human Error during Coordination Events

Human error is a normal artefact of work environments. In most cases the consequences of error are inconsequential; however, in some cases errors link together to create accidents, injuries and catastrophic failures of systems or 'organisational accidents' (Reason 1997). To manage error within a system a sophisticated approach involves addressing both error minimisation and error recovery/management.

After his early work on error typologies, Reason (1990) identified that organisations have in place a series of layers to protect themselves from the consequences of human error. He described four levels of human failure, each linked with the next. Training approaches can be seen to influence all these layers of protection:

1. organisational influences (e.g., training in leadership, resource allocation and culture)
2. supervision (e.g., training in management and decision making)

3. preconditions for Unsafe Acts (e.g., training in technological systems; in fatigue management)
4. Unsafe Acts (e.g., training in policies, procedures and decision making).

Given the ubiquitous nature of human error it is important to understand how it occurs within an emergency management coordination environment, especially if training systems are to be developed to help reduce and manage it.

With respect to this, Brooks et al. (2013) investigated the frequency and distribution of human error associated with selected major bushfire events across Australia. This study applied the Human Factors Accident and Classification System (HFACS) to examine three secondary sources of information associated with the Wangary, Canberra Fire Storm and Black Saturday fires. The sources are Commissions of Inquiry reports. The study identified a number of key conclusions:

1. Decision errors were often associated with time constraints, uncertainty, fatigue, the complexity of the situation, and personal interactions. At higher levels of coordination, decisions were sometimes made without questioning the veracity of the information, or with limited information or decision processes. This is consistent with the findings of Fallesen et al. (1996) regarding decision making errors in the US military.
2. Within the Human Factors Accident and Classification System (HFACS), skill-based errors are associated with skills that occur 'without significant conscious thought' (e.g., turning a steering wheel in a car). Coordinators of wildfire events, especially at the higher levels of operation, are rarely performing skill-based operations. This may require a reconceptualisation of skill-based errors in this domain.
3. Crew Resource Management issues such as failures in teamwork, communications or leadership constituted approximately 25 per cent of all issues across the three fires at the Incident Management Team level, suggesting that coordination within and between teams can be significantly improved. Emergency management competencies indicate the need for a range of personality-driven, technical and non-technical skills to support communication. This is discussed further in this chapter under the section on emergency management competencies.
4. Above the Incident Management Team level it is clear that Regional and State Control Centres struggled to effectively supervise Incident Management Teams. This reflects the challenges of supervising while allowing Incident Management Teams to manage their span of control, and is also a product of the complexity inherent in multi-team systems. Span of control is a concept that relates to the number of individuals or functional teams/groups that can be effectively supervised by one person, however simple.
5. The system of emergency management coordination is regularly degraded during a fire event. Whether this is due to a lack of information on the fire,

a lack of resources, or fatigue, those coordinating the response must often work outside the boundaries of what might be considered the 'safe system'.

6. The need to operate a degraded system and the sheer complexity of the emergency management effort suggests that managers need to be able to apply simple, robust 'heuristics' (rules of thumb) to manage emergency events.

7. Safety theory (e.g., Rasmussen 1997) suggests that one way to approach this is by making boundaries between safe, nearly safe and unsafe systems visible. This is different to traditional approaches that place 'defences' along pre-planned paths.

A summary of the results of this study can be found in Table 9.1. The results suggest that when emergency management and coordination 'goes wrong' in out-of-scale bushfire events this occurs because communication breaks down, poor decisions get made for a variety of reasons and those decisions are not subject to the sort of scrutiny necessary to 'catch' and correct them (see Owen, Chapter 7, this volume for further discussion on the importance of effective communication and high reliability). Coordination of these sorts of out-of-scale events also occurs in a resource constrained environment where the organisational processes/system sometimes cannot adjust to the scale of the event. The complexity of the situation is important in this respect.

Table 9.1   Distribution of human factors issues all control levels combined

All Control Levels *n = 118* *	N	%
Unsafe Acts (Decision-errors)	19	17
Unsafe Acts (Exceptional Violations of Procedures)	17	14
Preconditions (Crew Resource Management–Communication Issues)	28	24
Unsafe Supervision (Inadequate Supervision and Planned Inappropriate Actions)	34	29
Organisational (Resource Management and Organisational Processes)	10	8

* that only main categories of error are included and will not necessarily sum to the total number of errors.

Just like the layers of Reason's (1990) 'Swiss-cheese' model of accident causation, errors at different levels of coordination do not occur in isolation to each other. The effect of errors at a regional coordination level can have flow-on effects for the management of an Incident Management Team or a State Control Centre. If we take the Wangary fire as an example (see Figure 9.1), the interaction between errors can be demonstrated to occur both within teams and across teams. This fire occurred in January 2005 near Wangary, South Australia. The fire burnt over 77,000

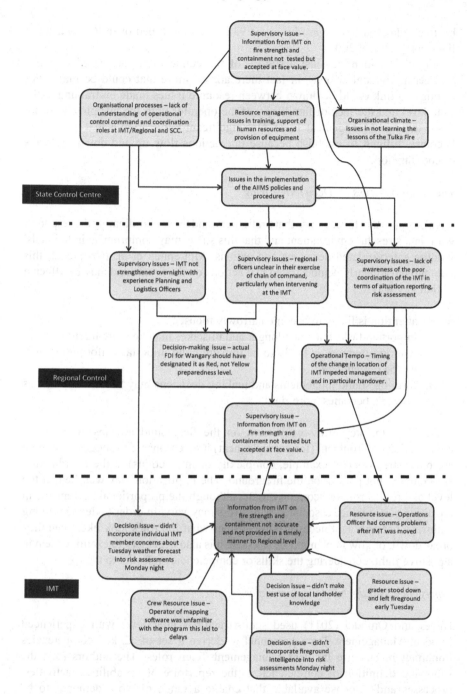

Figure 9.1   Interaction of errors – Wangary fire

hectares, claimed nine lives, destroyed 93 homes and killed over 46,000 head of livestock (Schapel 2006).

Figure 9.1 identifies just a sub-set of the connections that could be made between individual errors – in fact there are far more that could be made. For example, a link could be drawn between resource issues made on training at the state level and the lack of familiarity of the operator with the mapping software in the Incident Management Team. One significant implication is that 'causation' becomes difficult to establish because interactions flow up and down the paths of coordination.

The Effect of Stress on Decision Making

It is also important to recognise that firefighters, and those coordinating the response, work in a stressful environment and that this stress may compromise individuals' wildfire safety-related decisions and actions. McLennan et al. (Chapter 2, this volume) have clearly demonstrated that actions on the fireground may be affected in four key ways:

1. attention is likely to become narrowly focused
2. important tasks may take longer and mistakes may be more likely
3. working memory is likely to be impaired and important information may not be remembered
4. forming sound judgements and making decisions may become difficult as thinking becomes more rigid.

Although some key stressors present on the fireground are absent from the coordination environment (for example, heat) it is reasonable to suggest that there are other stressors (for example, complexity of interactions) in the coordination environment not present on the fireground. The implication may well be that the level of stress is roughly commensurate, although the proportional decrements in cognitive performance associated with stress may vary. In order to define training pathways not only is it necessary to consider what job roles look like when they break down, or how they might be stressed, it is also necessary to examine them in a positive light, considering the skills or competencies relevant to the role.

Emergency Management Competencies

Hayes and Omodei (2011) used semi-structured interviews with experienced Incident Management Team personnel to derive a set of 12 key competencies important in bushfire Incident Management Team roles. The authors used the following definition of competencies: 'the repertoire of capabilities, activities, processes and response available that enable a range of work demands to be met more effectively by some people than by others' (p. 3). Although there are elements of the coordination task above the Incident Management Team level that are different to working within the Incident Management Team, it would be still reasonable to assume some overlap between the competencies at these levels.

As indicated earlier in this chapter those competencies can be categorised in a number of ways; however, a basic delineation might be made between non-technical and technical skills (Flin, O'Connor and Crichton 2008), which also relate to task-work competencies and teamwork competencies (McIntyre and Salas 1995).

The list of key competencies from Hayes and Omodei (2011) is reproduced in Table 9.2.

Table 9.2   Key emergency management competencies

Key Competencies
Interpersonal and communication skills
Disciplined
AIIMS knowledge and processes
Management skills
Leadership
Decision making ability
Flexible and adaptable
Analytical thinking and problem solving
Calm and level-headed
Situational awareness
Technical expertise
Other (maintain sense of humour, self-confidence, demonstrate initiative)

The table can also be seen to identify some key 'groupings' of competencies, acknowledging a degree of overlap between these. Technical expertise and Australasian Inter-Agency Incident Management System (AIIMS) knowledge and processes might be considered technical or task related competencies. See Chapter 1 in this volume for more information about AIIMS.

Being disciplined, flexible and adaptable, calm and level-headed, self-confident, maintaining a sense of humour and demonstrating initiative are essentially qualities of the individual, strongly linked to personality type, and also correlated with expertise and experience. Finally, there are a range of non-technical or teamwork competencies that include communication/management skills, leadership, decision making and an ability to maintain situational awareness. This interaction is shown in Figure 9.2.

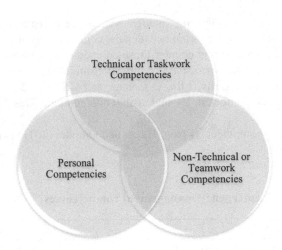

Figure 9.2   Competency types in emergency management

The categorisation of technical, personal and non-technical competencies is important for a number of reasons. Each category requires a different type of competency; however, the application of these types of competency must coexist during emergency events. The competency may also be more or less difficult to acquire. Competencies linked to personality-type may be very difficult for some to develop, and almost 'innate' for others. It is also possible that individuals working at regional or state levels (and therefore above Incident Management Teams) have 'evolved' to have these competencies because it appears that those individuals who do have these competencies tend to survive best in that environment.

An obvious question then becomes, how do we train people to improve taskwork, teamwork and personal competencies? Are the approaches for each of this group of skills different? We have already suggested that personal competencies are related to personality type and to the level of expertise, and to some extent are naturally selected by the organisational environment. However, studies of personality in occupations indicate that the distribution of personality types (based on profiles such as Myers-Briggs) in occupations is not significantly different from a random sample within a population (Pittenger 1993). What are the implications for training with respect to Hayes and Omodei's (2011) 'calm and level-headed' or 'disciplined' competencies from this perspective? Part of the solution may be to detach these competencies from personality profiles and instead focus on provision of experience through simulation and exercising to train control of emotional responses to highly novel and disturbing situations. For further discussion on the importance of managing emotions see Douglas (Chapter 5, this volume). This is a difficult and complex area and beyond the scope of the current chapter to examine in any significant depth. Instead we turn our attention to those teamwork skills so important to emergency management coordination.

Training to Enhance Non-Technical Skills

Other work domains have been training in non-technical skills for many years (Flin et al. 2008). Typically these approaches have followed the path of traditional training development, which is, performing a training needs assessment, developing the training materials and building tools to assess/evaluate the outcomes. Various strategies are available to train people in the development of non-technical skills. Approaches to Crew Resource Management (CRM), for example, have evolved since being introduced in the 1980s. Topics in Crew Resource Management training are designed to 'target knowledge, skills and abilities as well as mental attitudes and motives related to cognitive processes and interpersonal relationships' (Flin et al. 2008, p. 248), embedded in three distinct phases:

1. The Awareness Phase – this is the classroom component that introduces the theoretical aspects of non-technical skills.
2. The Practice and Feedback Phase – this is typically the simulation component; however, training strategies such as role-playing can also be used.
3. The Continual Reinforcement Loop – this includes refresher training and more recently the alignment of workplace auditing, standard operating procedures and training systems in order to reinforce non-technical skills through different aspects of the management system. In the absence of regular training and reinforcement, attitudes and practices tend to decay, so this also typically includes refresher training.

A number of training strategies are possible and Flin et al. (2008, p. 251) have identified when each of those strategies might be most appropriate (see Table 9.3).

It is important to note that these training strategies have developed to adopt a more holistic approach, recognising that unless training outcomes are also assessed in the workplace, and competencies are consistent with policies and procedures, performance will still not meet goals in terms of safety and reliability (Di Lieto and Brooks 2012).

Rasmussen (1997), in discussing the 'safety space' and the boundaries between safe and unsafe behaviour, provides a clue as to how we can train for these sorts of situations. 'Rather than striving to control behaviour by fighting deviations from a particular pre-planned path, the focus should be on the control of behaviour by making the boundaries explicit and known and by giving opportunities to develop coping skills at boundaries' (p. 191). Much of the previous research about identifying boundaries in the fire domain has been framed in the context of decision making. It is to this issue I now turn.

Table 9.3   Training strategies for developing non-technical skills

Type of Training	Specific Recommendations
Team coordination training	• Effective even with teams that do not have a fixed set of personnel • The training incorporates and emphasises non-technical skills • Currently applied to varying degrees within EM training and exercising in Australasia (Brooks and Owen 2013). Could be improved through further development of support tools such as guidance for building shared mental model during After Action Reviews
Cross-training	• Team has high levels of interdependence between members • There is a lack of knowledge about the roles of other team members • Good for high staff turnover environments or environments where teams are rapidly formed and changed over the period of the emergency • Good for liaison officers and those working in Control/Coordination Centres
Team self-correction training	• Team has high levels of interdependence between members • Good for low staff turnover environments • Could be applied to Incident Management Teams
Event-based training	• Useful when there are problems with a particular subset of tasks, and the tasks can be simulated • Currently applied to varying degrees within EM training and exercising in Australasia (Brooks and Owen 2013)
Team facilitation training	• Good when there are limitations in training resources • Could be supplemented with low-fidelity simulation

Developing Expertise in Decision Making

The history of decision making can be mapped to show a growing realisation that humans are far from perfect in their approach and in the outcomes of the decisions that they make. We don't evaluate all the alternatives, we don't comprehensively measure our utility and we are subject to a range of biases that lead us into error. These errors usually seem completely plausible at the time the decision is made. The corollary of this statement is that our ability to approximate, more often than not, allows us to make decisions that are reasonably accurate, or at least avoid major consequences most of the time.

However, in high risk situations, being 'reasonably correct' can be problematic – and evidence of decision errors during major bushfire coordination events has been found in earlier studies of secondary source material associated with major fire events (Brooks et al. 2013). This leads to several questions: Can we train people to improve their decision making during emergency management coordination events? What approach is best? How, if at all, should this be integrated with other training activities?

The Recognition-Primed Decision Making Model (RPD Model) was developed following observational research on firefighter decision making (Klein 1998). Klein noted that these commanders didn't seem to 'make decisions' as such – they just acted. Underlying these actions was an intuitive approach to decision making that had developed in response to the uncertain, dynamic environment that is urban firefighting. The model has three variants:

1. if the situation is a routine one and familiar to the commander, the decision is intuitive and automatic;
2. if the situation is not straightforward, but is ambiguous and unfamiliar, the commander must actively work at generating an accurate assessment of the situation, but once they have that assessment the action to take becomes obvious;
3. if the commander recognises a situation, but retains a degree of uncertainty over the appropriate action, they will envision an appropriate course of action and use mental simulation to mentally 'test' its effectiveness.

Decision Skills Training

If we are to train people in some way to make better decisions when managing emergency events, how should this occur? Klein (1997) suggested that this does NOT involve teaching specific decision strategies. There is little empirical research to suggest that trainers can identify superior decision strategies that could be used under 'field conditions' (i.e., in the control centre). Instead Klein (1997) suggested we should consider two different strategies to improve the expertise of the decision maker:

1. teaching people to 'think' like experts
2. teaching people to 'learn' like experts.

Klein's earlier work in this area involved revision of the firefighting training manuals for the National Emergency Training Centre in Maryland USA. He noted that 'what the instructional manuals seemed to need the most was an explication of the critical cues and judgements, so that readers could learn how more experienced commanders saw the world, and could thereby make progress in thinking (and seeing) as an expert' (Klein 1997, p. 347).

He identified the cues by working with an experienced retired commander and embedded the decision training within the existing course structure. There are, however, problems with this approach. Identification of cues, patterns and associated strategies can be complex and teaching them can be labour intensive; those cues may not be exhaustive for all situations, especially in the dynamic environment of emergency management coordination. Such problems led Klein to consider whether we might be able to train people to 'learn' like experts. 'We can provide tools for helping people gain expertise on their own, without trying to predefine the nature of that expertise' (1997, p. 347). He suggested that these tools included:

1. using attentional control exercises to practise flexibility in scanning situations
2. compiling an extensive experience bank
3. enriching experiences via review to derive lessons learnt and identify mistakes.

There is some evidence that many of the strategies identified by Klein (1997) are already being implemented within the emergency management domain to a greater or lesser degree. For example, the various organisations that have initiated 'Lessons Learnt' processes, in the Australian fire agency domain, could be considered an approach for compiling an extensive experience bank while deriving lessons learnt/identifying mistakes. There is also support for human factors training provided by the Australasian Fire and Emergency Authorities Council (AFAC) that supports non-technical communication, decision making and teamwork skills.

Thinking Critically and Practically

Critical thinking has been described as 'active, persistent and careful consideration of a belief or supposed form of knowledge in light of the grounds which support it and the further conclusions for which it tends' (Kiltz 2009, p. 9). This approach has been used in military environments to train naval officers (Klein 1997). The typology of Probing Questions has been applied in domains such as higher education to improve critical thinking skills. It identifies five question types:

1. clarification
2. assumptions
3. reasons and evidence
4. viewpoints or perspectives
5. implications and consequences.

Kiltz (2009) suggested that 'this approach provides infinite opportunities for critical thinking' (p. 14). Critical thinking can be aligned with the concept of 'Practical thinking' as applied in the (command and control) C2 environment of the US military (Fallesen et al. 1996). Practical thinking is based on a number of propositions:

1. thinking skills can be improved
2. thinking is not always positively correlated with IQ
3. reasoning errors can be decreased
4. thinking is goal-directed and done in context
5. models of normative decision making (where all alternatives are thoroughly evaluated and assessed for utility) and rules of formal logic are not very useful for improving practical thinking
6. recognise that not everyone thinks the same way, make people conscious of strengths, weaknesses of their personal thinking style.

'Practical thinking was defined...to consist of the application of creative and critical thinking skills to reason and reach conclusions about "everyday" situations and problems' (Fallesen 1996 et al. p. 36). The topics in this practical thinking course included the following elements:

1. critical thinking introduction, creative thinking
2. multiple perspectives (i.e., when and how to recognise)
3. metacognition (decision triage, crisis decision making, reorganisation, the value of concepts)
4. hidden assumptions (what else, detailed exercising, managing unexpected events)
5. practical reasoning in the face of uncertainty
6. integrative thinking
7. visualisation and prediction
8. diagnostics (before and after testing)
9. review and assessment.

Such a course/workshop would support the development in decision making skills likely to be required above the Incident Management Team level in out-of-scale events.

Developing Heuristics and Managing Biases in Emergency Management

In other high risk domains such as maritime transportation, rail and aviation, it has been necessary to arm people with effective 'quick decision-tools' that are cognisant of that environment. They must be able to be used dynamically, they must be relatively straightforward and they must support the development of shared mental models of the current situation and a prediction of future states (i.e., build situational awareness).

Biases (preferences for certain decision approaches) and heuristics (rules of thumb) are two characteristics of decision making that impinge on emergency management and must be acknowledged in any enhanced training systems.

At least since the work of Kahnemann, Slovic and Tversky (1982) we have recognised the existence of a range of biases when making decisions. People may suffer from a representative bias, for example, making generalisations from a small sample size. Wilson et al. (2007) tested the prevalence of risk-based biases in 206 line officers and incident command personnel from the USDA Forest Service. They concluded that the participants exhibited biases such as:

1. loss aversion (choosing a safer option when the consequences of the choice were framed as a potential gain)
2. discounting (choosing to discount short-term over longer-term risk in the belief that this longer-term risk can be controlled).

Such biases do not necessarily lead to decision errors – choosing safer options and discounting short-term risks are not inherently incorrect decisions. However, biases are just that – a preference for a certain decision result which can lead a decision maker to ignore contradictory evidence. Biases tend to reduce a decision maker's flexibility.

Heuristics (sometimes called rules of thumb) are strategies using readily accessible information to control problem solving. They are typically general in nature such that they can accommodate many situations, and rely on the notion that approximate action will be good enough. This is pertinent in situations that are dynamic, complex and uncertain – all characteristics of emergency management coordination.

The Zone of Coping Ugly (ZOCU)

A response is needed to Klein's (1997) suggestion that we should teach people how to build mental models and practise flexibility in scanning a situation. Coupled with this, it is also necessary to recognise that coordination of emergency events is often about managing a system that is 'degraded' in many ways – through a lack of resources, or information or because of fatigue.

A useful way to understand this 'degraded operation' might be to consider that there is an 'area' of functional performance (Rasmussen 1997) available within any system and boundaries at which errors can be made and recovered from. Outside these boundaries error recovery becomes either less likely or unlikely and adverse outcomes occur.

The work of Bonanno (Bonanno 2004; Bonanno et al. 2011) suggested that we can place degraded operations in a more familiar context. This approach may be to understand when we are 'coping ugly'. This approach might be defined as instances where we use strategies that wouldn't normally be considered reasonable but are a necessary part of operating in a degraded system. More importantly, because components of systems are bound to fail in such extreme coordination events, this might allow other parts of the coordination structure to recognise and adjust their own actions and potentially assist those components of the system to find their way back within the boundaries of what might considered safe or at least manageable.

In safety management systems, numerous precautions are taken to protect workers against occupational risk and the system against major accidents. Commonly we apply a 'defence-in-depth' design strategy to catch errors or issues and resolve them.

While the role of in-depth defences is valuable, when they fail consecutively or in such a way that insufficient defence remains, an accident becomes likely. When systems are stressed – because of production pressures or time constraints or simply the magnitude of the response effort (such as occurred in Australia with the Victorian 'Black Saturday Fires' in 2009) – it is likely that individuals and groups operated for significant amounts of time with degraded or insufficient defences. With respect to this issue, Rasmussen (1997) suggested the following solution: 'Rather than striving to control behaviour by fighting deviations from a particular pre-planned path, the focus should be on the control of behaviour by making the boundaries explicit and known and by giving opportunities to develop coping skills at boundaries' (p. 191).

Perhaps emergency management is similar:

> An important attribute of expert decision makers is that they seek a course of action that is workable, but not necessarily the best or optimal decision...time pressures often dictate that the situation be resolved as quickly as possible. Therefore it is not important for the course of action to be the best one; it only needs to be. (Phillips, Klein and Sieck 2004, p. 305)

Figure 9.3 depicts these concepts spatially and identifies the Zone of Coping Ugly. The white arrow shows the drift from safe to less safe performance to situations that may include accidents and incidents. These are the boundaries that Rasmussen (1997) talked about and approaching or crossing them indicates a need to adjust the coping strategies being used.

One of the most significant problems in thinking and decision making for out-of-scale events therefore occurs when people lose flexibility and creativity, and instead become anchored to a particular solution. This leads to the conclusion of the need for a 'playbook' of coping repertoires. Coping repertoires are groups of coping responses derived to deal with certain situations. They are our plans, as well as our Plan Bs, Cs, Ds etc...and plans that emerge organically in response to changing events. As Perlin and Schooler noted (1978), 'perhaps effective coping depends not only on what we do but also on how much we do' (p. 14).

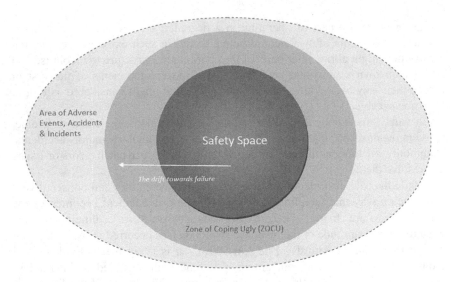

Figure 9.3   The safety space and the Zone of Coping Ugly

Within the Victorian Bushfires Royal Commission there is perhaps an eloquent example of the issue associated with the coordination of degraded systems and the concept of 'coping ugly':

> Concerning the issue of the Kangaroo Ground ICC being prevented from releasing warnings, Mr Rees went on to say that there was a need for flexibility and 'we...weren't as flexible as what we should have been'. Mr Rees said you have to give people 'the capability to break through [the system] when it is not working'. The Commission strongly supports this approach. (Teague, McLeod and Pascoe 2010)

In order to advocate for particular heuristics in the emergency management domain they need to be tested for their reliability and validity via training/exercising. This would create an evidence base that could then be fed into a non-technical skill programme following the three key stages of awareness raising, practice and continual reinforcement.

Implications for Practitioners and Instructors Earlier, it was noted that in Australasia current training systems application of human factors skills and knowledge are neither universally poor, nor systematically complete. The reality is that they sit somewhere in between these extremes. In order to demonstrate this it is necessary to look at training in its different contexts, that is, formal training and higher education, continuing professional development courses and operational training via the Australasian Inter-Service Incident Management System (AIIMS).

Formal Training Programmes

In Australia, the formal training for roles at, below and above the Incident Management Team level include non-technical skill training, building competencies required by firefighters involved in structural firefighting operations. The core units contained within Certificates II and III in Firefighting Operations reflect the competencies required by firefighters involved in wildfire firefighting. Certificate IV covers leadership and supervisory functions, while the Diploma and Advanced Diploma qualifications address management functions. Within the various levels of the Qualifications Framework are units relevant to training for effective emergency management coordination. A simple first question to ask might be whether there exists a viable training pathway for human factors-related concepts and competencies. By way of example, we might consider the training pathway associated with working in, leading and managing teams as illustrated in Table 9.4.

Table 9.4   Training pathways for working in, leading and managing teams

Certificate II	
PUATEA001A	Work in a team
Certificate IV	
PUATEA003A	Lead, manage and develop teams
BSXFMI402A	Provide leadership in the workplace
Diploma	
PUAOPE005A	Manage a multi-team response
PUAOPE001A	Supervise response
PUAOPE004A	Conduct briefings/debriefings
PUAOPE005A	Manage a multi-team response
BSXFMI506A	Manage workplace information
PUAOPE007A	Command agency personnel within a multi-agency emergency response
PUACOM007A	Liaise with other organisations
Advanced Diploma	
PUAOPE008A	Coordinate resources within a multi-agency emergency response
PUAMAN003A	Manage human resources
Graduate Certificate*	
S1	Foundations of Strategic Management
S3	Leadership and Organisational Change
Graduate Diploma*	
S1	Contemporary Leadership
S2	Executive Leadership Development

* As delivered by Australian Institute of Police Management

I might, therefore, suggest that if an individual had completed this 'pathway' they would have an effective understanding of the theory and practice associated with managing teams. One issue might therefore be how many personnel working in leadership positions in coordination environments at Regional and State Control Centres have these qualifications.

Informal Training Programmes

In the discussion paper *Professional Development Pathways for Protecting Fire Fighters, Human Factors and Emergency Management Leadership*, Owen and Omodei (2009) considered how research from the Bushfire Cooperative Research Centre (BCRC) was translating into education and training outcomes. The authors reviewed key projects on health research, decision making and incident management teamwork. They noted the apparent demand for professional development and informal courses in these areas, in association with formal models such as that offered by Australian and New Zealand Schools of Government or the Australian Institute of Police Management. This begins to build a picture that supports a multi-dimensional approach to training providing both formal and informal opportunities.

A further BCRC report identified the scope of current continuing professional development courses in this area, plus a suite of intermediate and advanced courses currently not delivered but which would be integrated into a holistic approach to human factors training within the emergency management space. The proposal was 'aimed at assisting agencies to build capacity in understanding those human factors that are currently believed to influence decision making and performance in emergency services contexts' (Krusel, n.d. p. 2). The author linked learning and development, safety management systems and operations via changes in culture, communication and thinking processes directed towards the principles of high reliability.

Operational Training in Human Factors

Associated with this, we might ask how effective the training is translated into practice, the assessment of that practice, and the relationship with agency training and exercising. It is therefore not enough to identify that formal training pathways exist to improve emergency management coordination.

To assess this we performed a desk-top review of the Australasian Inter-Service Incident Management System (AIIMS) training documentation (including Facilitator's Guide, Participant Workbook and PowerPoint presentations) for the modules listed in Table 9.5. The review indicated that a significant proportion of the fundamental human factors issues and skills were covered in these modules for activities at the Incident Management Team level.

Table 9.5   AIIMS training in human factor issues

S2	Human Factors and Incident Management
S12	Incident Action Planning
S13	Communication
S14	Incident Information
S15	More Thinking
S16	Leadership and Teamwork

For this material to translate to above the Incident Management Team level it would need to include further layers of complexity (i.e., the concepts in the modules have been somewhat simplified to suit the target audience). Beyond this, however, it will be necessary to define how the functional differences of working above the Incident Management Team (i.e., in terms of complexity, strategic as opposed to operational) alters the type of skills and knowledge coordinators need. At present it is not possible to define this systematically, given the different ways different jurisdictions manage and organise above the Incident Management Team.

Is the Australasian Emergency Management domain ready for the 'Resource Management' training approach?
Formal training, operational training (via AIIMS) and informal training approaches currently available indicate that there are training pathways for emergency management coordinators to achieve some level of competency associated with human factors issues. It would also, however, be reasonable to suggest that coverage could be more comprehensive and integrated. It may be that elements of these courses should be folded into a version of Crew Resource Management (CRM), perhaps Emergency Coordination Resource Management (ECRM). There are some further activities that will need to be completed in order to underpin such a training product. Importantly the training will need to consider the three phases of non-technical skill workshops outlined earlier – linking classroom/awareness raising with exercising and continual reinforcement of the concepts and associated skills.

Lowe, Hayward and Dalton (2007), in discussing the implementation of Rail Resource Management (RRM), noted that organisations with proactive or generative safety cultures will find it easier to implement these sorts of programmes, and will see greatest effect from them. However 'in any organisation...RRM training can contribute to the development and maintenance of a positive safety culture' (p. 27).

Perhaps more importantly the basic research that underpins a resource management style training approach needs to have been conducted. In this respect, under the auspices of the Australasian Bushfire Cooperative Research Centre, investigations of leadership, culture, decision making, heuristics and biases, human error, and breakdowns and disconnects have been completed or are currently being finalised, as demonstrated by the many publications cited in this text.

We considered this current research output as well as frameworks of courses in maritime (Swedish Club 2012), aviation (Australasian Civil Aviation Academy 2013) and rail (Lowe et al. 2007) transportation domains to derive a broad structure for an Emergency Coordination Resource Management course. The biggest surprise from reviewing resource management courses in other domains was the level of alignment between domains. Less than 40 per cent of unit descriptions could be considered the same across all three domains; another 40 per cent were shared by two of the three courses. Given this, there is a need to individualise resource management courses for the emergency management domain. Table 9.6 outlines a structure for an Emergency Coordination Resource Management course, with brief explanations of the relevant modules.

Table 9.6 Indicative components of an Emergency Coordination Resource Management course

M1 **Situational Awareness (SA)** – Recognition of multiple perspectives/models of SA; the interaction between this concept and mental models/shared mental models.
M2 **Communications, Briefings and Media** – Given the complexity and duration of EM events as well as the distribution of teams these issues deserve a major focus, including how communication disconnects and breaks down.
M3 **Culture** – EM organisations work across different organisational, occupational and national cultures; this module explores cultural typologies and their implications.
M4 **Human Error and Breakdowns** – Understanding types of error, error chains, error management, contexts in which error becomes more likely and remedial actions to minimise them. Identifying types of communication breakdowns and typical strategies for resolution.
M5 **Decision Making and Sense Making** – Explores decision making processes, effects of stress on these processes, and situations in which sense making may be a viable alternative.
M5 **Heuristics and Biases** – Applies rules of thumb to coordination contexts and looks at how biases may affect the process of coordination.
M7 **Workload, Coordination and Control** – Explores mechanisms for adjusting workload and identifying stress as a result of excessive workload. Establishes functional differences between coordination/control paradigms and implications.
M8 **Risk Management** – Explores concepts of risk/reward and application of simple risk matrices to quickly assess risks during EM events.
M9 **Leadership and Managing Up** – Explores modes of leadership relevant to the EM domain and how the concept of transformational leadership can facilitate change and growth in organisations, looks at strategies for managing the interface between political and operational environments.
M10 **Teamwork, Assessing/Managing Self and Others** – Investigates the key aspects of managing span of control and how knowledge of and strategies for emotional intelligence can be used to assess and manage oneself and others.

An Emergency Coordination Resource Management course is not, however, a complete training pathway for managers working above the Incident Management Team level. A complete pathway links formal training and education with continual professional development and agency or multi-agency based training and exercising. In its most flexible application it may allow individuals to 'learn like experts'.

Courses in this area have promoted the development of specific skills through the provision of understanding and by developing competency in the use of particular strategies to enhance non-technical elements such as communication, leadership and situational awareness. More recent courses in the maritime domain have supplemented this more traditional material with specific training on the challenge of managing increasingly autonomous systems and the implications this has for bridge team's situational awareness.

In Figure 9.4 building on the concepts identified by Krusel (n.d.) we have linked the sometimes disconnected elements of learning and development, operations and safety systems, and strategic leadership and organisational change. The original figure (Krusel, n.d.) created an excellent visualisation of what an emergency management organisation would need to address in order to achieve

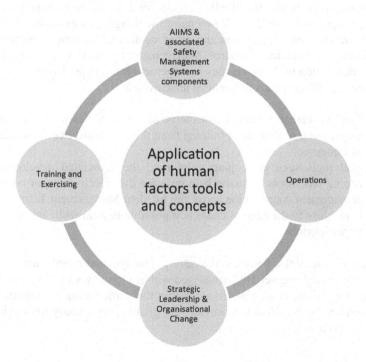

Figure 9.4   Enhancing EM capabilities through the application of human factors

high reliability. Importantly, the 'glue' includes managing change at a number of levels and the figure places this at the base of the cycle to demonstrate the importance of change management in this context. Perhaps the most significant challenge is achieving consistency between these elements. The obvious starting point is to settle the strategic vision and then build the change management plan for operation, training and the management systems in parallel.

Conclusions

Emergency management coordination faces challenges that are in some respects similar and in some respects different to other work domains. The non-technical skills such as leadership and communication are necessary irrespective of the domain; however, the complexity of the work environment, the distributed nature of the teams and the associated decision making, and the naturally eroded safety management system are just three examples of why approaches adopted in other domains cannot simply be applied within emergency management.

A review of Australasian emergency management human factors research, such as the work on decision making and stress (see McLennan et al., Chapter 2, this volume) reveals that the building blocks for a 'resource management' style training programme now exist. When paired with agencies expertise the design of ecologically valid simulations and other contextual training approaches, an Emergency Coordination Resource Management (ECRM) course could be a significant addition to the emergency management training landscape.

There are still questions that remain unanswered:

1. Can we actually teach non-experts to think or learn like experts? A longitudinal evaluation of training from this perspective could help answer this question.
2. Will tools such as 'coping ugly' (and the associated development of coping repertoires) actually help improve the effectiveness of emergency management coordination above the Incident Management Team?
3. How close to the edge of chaos is it possible to train while doing no harm to participants?

Importantly, the links between the safety management/coordination system, operations and training need to be managed within a framework that acknowledges the need for change at multiple levels within the participating organisations. An integrated approach such as this provides the greatest opportunity for maximising training outcomes.

Acknowledgements

The research was supported by the Bushfire Cooperative Research Centre Extension Grant. However, the views expressed are those of the author and do not necessarily reflect the views of the Board of the funding agency.

References

Australasian Civil Aviation Academy (2013). *Crew Resource Management course structure* accessed on 24/2/13 from http://www.caaa.com.au/training/crew_ resource_management/

Bonanno, G. (2004). Loss, trauma and human resilience. *American Psychologist*, 59(1), pp. 20–28.

Bonanno, G., Westphal, M. et al. (2011). Resilience to loss and potential trauma. *Annual Reviews in Clinical Psychology*, 7, pp. 1–25.

Fallesen, J., Michel, R. et al. (1996). *Practical thinking: Innovation in battle command instruction*. Technical Report 1037. Alexandria: U.S Army Research Institute.

Brooks, B., Bearman, C. et al. (2013). Human error during complex bushfire coordination events: An analysis of Australian bushfires using the human factors analysis and classification system. *Safety Science*. Under review.

Brooks, B. and Owen, C. (2013) *Report on training pathways in human factors for the coordination of large-scale bushfire events*. Report prepared for the Bushfire Co-operative Research Centre, Melbourne.

Di Lieto, A. and Brooks, B. (2012). *The Maritime Operations Safety Analysis (MOSA)–Confidential Report*. Launceston: Australian Maritime College.

Flin, R., O'Connor, P. and Crichton, M. (2008). *Safety at the sharp end: A guide to non-technical skills*. Aldershot: Ashgate.

Hayes, P. and Omodei, M. (2011). Managing emergencies: Key competencies for incident management teams. *The Australian and New Zealand Journal of Organizational Psychology*, 4, pp. 1–10.

Kahneman, D., Slovic, P. and Tversky, A. (1982). *Judgement under uncertainty: Heuristics and biases*. Oxford: Oxford University Press.

Kiltz, L. (2009). Developing critical thinking skills in homeland security and emergency management courses. *Journal of Homeland Security and Emergency Management*, 6(1), pp. 1–23.

Klein, G. (1997). Developing expertise in decision making. *Thinking & Reasoning*, 3(4), pp. 337–352.

Klein, G. (1998). Naturalistic decision making. *Human Factors*, 50(3), pp. 456–460.

Krusel, N. n.d. *Enhancing capability through human factors professional development*. Melbourne: Bushfire CRC.

Lowe, A.R., Hayward, B.J. and Dalton, A.L. (2007). *Guidelines for rail resource management*. Brisbane: Rail Safety Regulators Panel.

McIntyre, R. and Salas, E. (1995). Measuring and managing for team performance: Emerging principles from complex environments. In R. Guzzo, E. Salas and Associates (eds), *Team effectiveness and decision making in organizations*. San Francisco: Jossey-Bass, pp. 9–45.

McLennan, J., Strickland, R. and Omodei, M. (2012). Wildfire safety-related decisions and actions: Lessons from stress and performance research. *IAWF Safety Summit, October 2012*. Sydney, pp. 33–46.

Murphy, P. and Dunn, P. (2012). *Senior leadership in times of crisis*. Deakin West, ACT: Noetic Group Ltd.

Owen, C. and Omodei, M. (2009). *Professional development pathways for protecting firefighters, human factors and emergency management leadership – a discussion paper*. Report prepared for the Bushfire Co-operative Research Centre, Melbourne.

Perlin, L. and Schooler, C. (1978). The structure of coping. *Journal of Health and Social Behaviour*, 19(1), pp. 2–21.

Pittenger, David J. (1993). Measuring the MBTI...and coming up short. *Journal of Career Planning and Employment*, 54(1), pp. 48–52.

Phillips, J.K., Klein, G. and Sieck, W.R. (2004). Expertise in judgment and decision making: A case for training intuitive decision skills. In D.J. Koehler and N. Harvey (eds), *Blackwell handbook of judgment and decision making*, pp. 297–315.

Rasmussen, J. (1997). Risk management in a dynamic society: A modelling problem. *Safety Science,* 27(2/3), pp. 183–213.

Renaud, C. (2010). *Making sense in the edge of chaos: A framework for effective initial response efforts to large scale incidents*. M.Arts Thesis. Monteray, Naval Postgraduate School.

Reason, J. (1990). *Human error*. Cambridge: Cambridge University Press.

Reason, J. (1997). *Managing the risk of organizational accidents*. Aldershot: Ashgate.

Schapel, A. (2006). *Inquest into the deaths of star Ellen Borlase, Jack Morley Borlase, Helen Kald Castle, Judith Maud Griffith, Jody Maria Kay, Graham Joseph Russell, Zoe Russell-Kay, Trent Alan Murnane and Neil George Richardson*. Coroners Court of South Australia, Adelaide.

Swedish Club, (2012). *Maritime Resource Management course structure*. Accessed on 24/2/13 from http://www.tscacademy.com/main.php?mcid=3&mid=166&pid=17246&tid=17245.

Teague, B., McLeod, R. and Pascoe, S. (2010). *2009 Victorian Bushfires Royal Commission final report*. Melbourne: VBRC.

Wilson, K.A., Salas, E., et al. (2007). Errors in the heat of battle: Taking a closer look at shared cognition breakdowns through teamwork. *Human Factors*, 49, pp. 243–256.

Chapter 10

Creating Cultures of Reflective Learning in the Emergency Services: Two Case Studies

Dr Sue Stack

Bushfire Cooperative Research Centre, Melbourne, Australia and University of Tasmania, Australia

'I valued the opportunity to reflect on a real world situation and look at what happened in some detail. We don't normally create the situation to have this reflection.'

(Cobaw Staff Ride participant – Fire Leader)

Why Reflective Learning is Important

Agencies involved in fire and natural hazard management experience a range of organisational and operational conditions that provide considerable challenges. Incidents are highly complex, high risk, with ill-defined or changing goals, often with multiple goals in conflict. There is information overload and few things can be manipulated or controlled (Klein et al. 2003).

There is a mismatch between the limited capacity of the human mind and the complexity of managing emergency incidents such as bushfires (see for example Frye and Wearing, Chapter 4, this volume). The result of this mismatch is that cognitive biases or 'error traps' in thinking are highly likely, with often serious consequences for health, life and property (Dekker 2006; Reason 2008; Frye and Weaving 2011). Omodei (2012) suggested four core strategies that help individuals overcome the biases that lead to error traps: engaging in meta-cognition (awareness of the limitations of thinking and being able to regulate their own thinking), reflection, anticipatory thinking, and considering other perspectives. However, Reason's (2008) *Swiss Cheese model* highlights that while error might appear to be at the operational level, it is more likely to be a combination of many different organisational layers, including larger organisational issues and trade-offs (see Bremner, Bearman and Lawson, Chapter 8, and Brooks Chapter 9, this volume).

Weick and Sutcliffe (2001) suggested that organisations involved in such difficult domains need to have a whole culture approach to ensuring their field of operations is 'high reliability'. This culture is based on the notion of *mindfulness:*

1. a desire to look for and understand mistakes as weak signals alerting to systemic issues

2. reluctance to simplify as simplification will not capture the reality of what it means to work in uncertain, ambiguous and dynamic environments
3. being sensitive to the detail of operations because weak signals and cultural nuances will emerge
4. inviting those with expertise to speak out and inform decision making regardless of rank
5. building resilience of employees.

One key to mitigating against individual cognitive biases which can lead to organisational failures is engaging in reflection *in* action (during the performance of one's work) and *on* action (reflecting after the performance on what occurred) (Schon 1983). However, reflection *in* action is only possible if there is enough mental processing capacity at the time of the experience to get 'outside' the act of generating the performance to be able to both watch and evaluate. It is particularly difficult to do under degrading situations in the heat of action on a fireground or in an incident management team when things are very busy. According to Weick (2002), under such circumstances people might have reduced ability to interpret situations. They may revert to previously held command roles and unconsciously adopt ingrained routines (whether helpful or not).

Phillips, Klein and Seick (2004) highlighted the importance of reflection-on-action in building firefighter expertise. They suggested that careful study of a case with attention to key decision making points and their cues provides 'mental slides' that become part of a person's memory bank. In time-poor operational situations these cases are unconsciously drawn on through the process of *Recognition Primed Decision Making* where situational cues are matched with those in the case. The way experts build their capacities include reflective thinking about theirs or other's incidents and the seeking out of new cases to add to their memory (Phillips et al. 2004).

Developing Reflective Cultures

Reflection has long been considered an important part of the learning process, enabling experience to be reflected upon, generalised and then applied to new situations (Dewey 1933; Kolb and Fry, 1975). Boud, Cressey and Docherty (2006) described *productive reflection* as an activity that we choose to do in particular situations where 'ambiguity cannot be controlled and managed as a routine process, so reflection as an open process that deals with matters that by definition do not have a ready solution or are not clearly formulated is needed' (p. 22). The activity of reflection is quite distinct from analysis. 'Reflection is a discursive way of creating a space for focusing on problematic situations and holding them for consideration without a premature rush to judgement' (Boud et al. 2006, p. 23).

However, Kemmis (1985) warned that reflection is a problematic exercise. It has the potential to *reproduce* or to *transform* practices, relationships between people, decision making and communication. According to Schon (1983), a key potential of reflective activity is that it can help us consider or adopt new 'frames' (of sense making), thus helping us to move outside habitual thinking and behaviour patterns. A frame might be a 'mental model' or a perspective. It can be informed by a theory, a world view or a value system. Frames can be shaped by individual ways of being in the world, group ways of operating and thinking, and organisational cultures and ideologies. Reflection that does not include the consideration or experience of alternative frames can entrench us into existing thinking and behavioural patterns.

Yorks and Marsick (2000) suggested that the impact of reflection can be considered on a spectrum from in-form-ation to trans-form-ation, as shown in Table 10.1.

Table 10.1   Information to transformation

	Form	Process
In-form-ation	Elaboration *within* existing frames of reference	Gather content Incidental reflection
	Analysis *within* existing frames of reference	Reflection on content
	Adopting *new* points of view	Reflection on process and content
Trans-form-ation	Adopting *new* cognitive structures	Reflection on process, content and premises

Reflecting on *processes* as well as content provides the opportunity to adopt new points of view or perspectives through which information is interpreted in a different way. Reflecting on *assumptions* can provide the impetus for transformation to new cognitive structures where a person may develop a greater capacity to deal with complexity and be able to hold multiple perspectives and identities in dialogue (Kegan 1994; Taylor, Marienau and Fiddler 2000).

Hoyrup and Elkjaer (2006) suggested a useful framework in considering *productive reflection* within work contexts that include four perspectives: individualised, critical, socialised and organisational:

1. **Individualised perspective** – 'Reflection is prompted by a complex situation involving problems, uncertainty and ambiguity' (Hoyrup and Elkjaer 2006, p. 32). Reflection is primarily done by the individual on their experience, but may be assisted by others. Reflection might draw on different elements such as returning to the experience, defining the

problem, reframing, evaluating, problem solving, anticipatory thinking, testing, reconstructing knowledge, attending to feelings.

2. **Critical perspective** – Critical reflection is questioning the taken-for-granted or 'hunting assumptions', such as analysis of power relations within social, cultural and political dimensions (Brookfield 1995). This is associated with double loop learning (Agryis and Schon 1974), which may result in the movement to new values, perspectives, cognitive structures or action logics (Torbert 2004).

3. **Socialised perspective** – 'The process of reflection is collective, we reflect together with trusted others in the midst of practice' (Hoyrup and Elkjaer 2006, p. 36). Reflection is a commitment to inquire together through learning from mistakes, vision sharing, sharing knowledge, challenging group think, asking for feedback and experimentation. It could have the capacity to sense beyond the individual using collective presencing (Senge 2004; Scharmer 2009).

4. **Organisational perspective** – This is the organising process in order to create and sustain opportunities for organisational learning and change. This may be through developing communities of practice (Wenger and Lave 1991) and developing shared meta-language that enables people to describe what has been tacit (Stack and Bound 2012). This may include collective questioning of assumptions underpinning organisations, leading to change in structures or decision making. It may also be subject to abuse by management where participants are led to feel a sense of agency but nothing comes of it (Boud 2006).

For organisations considering developing reflective cultures all these layers become important: from developing individual capacity to providing opportunity and support for it to occur. It is useful to understand and acknowledge the practices for reflection that already exist within organisations. For example, the emergency services are often under high media scrutiny, reviews or litigation that can create blame-based cultures, which may cause burying of mistakes or witch hunts. Reflection may be associated with technical application of schemas or clinical reviews that do not provide the opportunity for double loop learning. The transformation of the organisation into one based on trust and a quality of reflective practice as described above is part of a trajectory over many years with the challenge to find opportunities that might provide models for the sort of culture that an organisation might wish to adopt.

The Staff Ride

The Staff Ride is a learning programme that can act as a bridge between reflection within an educational setting and reflecting at work. It is an immersive reflective experience which revisits an incident on the ground where it actually occurred (Sutton and Cook 2003) providing an opportunity for improved situated understanding (Eraut 2000). It is a programme that addresses many of the issues raised above; providing organisational support to encourage individual reflectivity and open and thoughtful dialogue within a non-judgmental social learning setting.

The Staff Ride originated in the military where officers would go over a battleground and review the sequence of action and decision making. Staff Rides have continued to be used by the military and are a currently a well-established programme of learning for the Fire Services in the USA, with many different formats and learning objectives, including hearing the stories of original actors at critical moments, or facilitating tactical decision making games around key decision points.[1] Their audience is now seen to be wider than the original participants engaged in an after-action review (post fire incident inquiry report on what happened); they may be conducted for an existing group such as a fire brigade, or provide an opportunity for people from different organisations and roles to reflect together on the experiences of others.

The Staff Ride presents a rich complex case study for learning, where it is often in the detail, in the nuances, in trying to walk in someone else's shoes that people are able to make insights or perspective shifts that impact on how they see things and act in the future. Shulman (1992) suggested that good case studies motivate, ground, avoid over-simplification or generalisation, provoke alternative readings of situations, use narrative to help get inside the personal motivations and conceptions of people, and capture the culture and the context in which the story appears. Shulman says that while learners may draw principles from cases, key outcomes might be a new flexibility in thinking. He also suggested that learning from case studies can reduce the problems often cited of transfer of learning because they can simulate the contexts that people work in, enabling situated cognition.

As case studies, Staff Rides usually have three phases:

1. Preliminary Study – participants are provided with pre-reading about the incident. Immediately prior to the field trip they may have an orientation to the event, key speakers, or introduction to relevant theories.
2. The Field Trip – visit to the site where participants move through different *stands* (locations) in sequence of the incident. This may or may not involve the original participants in the incident telling their story. It may involve discussion, tactical decision making exercises or role plays.
3. Integration – opportunity for participants to reflect and integrate what they have heard. This may involve an informal dinner or staying overnight with a morning session.

The Staff Ride aims to walk the tension of being non-blame as well as providing critical review:

---

1   See the Wildland Fire Lessons Learnt Centre at http://wildfirelessons.net/Home. aspx.

- Allowing participants in the original incident to tell their stories without blame about what they did, providing detail and openness about thinking processes so the listeners can begin to build up a shared meaning of the situation and an understanding of why decisions might have been made.
- Fostering critical reflection on what might have shaped those decisions – making visible implicit processes, assumptions, power relations, organisational cultures and structures that go beyond the individual.

The rest of this chapter aims to look at two programmes recently run in Australia. One programme, *The Narawntapu Staff Ride,* was developed by the Tasmanian land management agency, Tasmanian Parks and Wildlife Service (PWS), in 2010. The other, *The Cobaw Staff Ride,* was developed jointly by the Victorian land management agency, Department of Sustainability and Environment (DSE) and the Country Fire Association (CFA) in 2012. Both programmes aimed to build reflective cultures and practice within their organisations, drawing from the USA Wildfire Lessons Learnt programmes.

Method

The approach taken in this chapter is to use the two case studies to highlight particular learning related to reflective inquiry. The Narawntapu Staff Ride case study follows the journey of the organiser in developing the programme and then modifying it for further contexts. It examines the layers of reflection that became important in this process. It aims to give future organisers or facilitators of such Staff Rides a sense of some of the complexities in designing effective learning from a Staff Ride. The Cobaw Staff Ride case study describes an alternate approach to a Staff Ride, discusses the quality of reflection that occurred and provides some measures of its impact on the attendees (see Table 10.2 for a summary of the two rides).

The Narawntapu Staff Ride case study is based on my observation and videoing of the Narawntapu Staff Ride, interviews with participants, and subsequent conversations with the organiser Sandy Whight as we developed a guide for running Staff Rides, reflecting on how to improve the learning. Some of these learnings are captured in *Designing the Staff Ride* (Stack et al. 2010), which accompanies a 50-minute video of the Narawntapu Staff Ride, and a 10-minute Lessons Learnt video.

Table 10.2   Summary of the Narawntapu and Cobaw Staff Rides

Staff Ride	Length	Stands	Participants	Preliminary Study	Field Trip	Integration
Narawntapu	Overnight prior to field trip	7	14 leaders in fire management section	2-page burn plan, session on High Reliability Organisations the evening before	5.5 hours 1 speaker	30 mins at site
Cobaw	Overnight prior to field trip	6	40 participants for 2 occasions, targeted as a vertical slice of leadership in fire management	120-page booklet, orientation to the event and human factors the night before	7 hours, 4 facilitators, role plays, 1 speaker at last stand	30 mins at venue

The Cobaw Staff Ride was evaluated using Cafferella's (2002) integrated evaluation approach, which involved formative assessment of the preliminary design and engagement of stakeholders to enable improvement, followed by a summative evaluation of the programme drawing on the first three of Kirpatrick's (1998) four levels of evaluation (participants' reactions, learning and impact). The Cobaw Staff Ride programme involved two cohorts of 40 participants (n = 80) who were sent pre-reading about the event two weeks prior and required to complete a pre-Staff Ride survey. The pre-Staff Ride survey sought information about their experience with managing fire incidents, their previous training programmes and their perceived impact, their impressions from the pre-reading, their expectations for the Staff Ride and a self-assessment of the type of organisational culture participants thought best represented their agency. Participants completed a post-Staff Ride survey one to three weeks after the event that sought their insights following the programme, how their thinking had changed, whether they had changed their practice, what they valued about the programme, suggestions for improvement, and an indication of whether they would like to be involved in further Staff Ride experiences (as facilitators or using their own incidents). Both surveys used a combination of descriptive elements and Likert-type questions.

Quantitative comparisons using a related sample Wilcoxin Signed Rank Test were made between participants' experience of the Cobaw programme with other learning programmes they had experienced, vignettes were composed of participant responses to highlight key themes emerging from the findings. The first programme was observed by me, with interviews of participants, video and recordings, and discussions with facilitators and organisers. This captured some of the phenomenology of the experience (Stack and Owen 2012).

The Narawntapu Staff Ride – Through the Lens of a Journey

This case study follows the learning journey of Sandy Whight, Planning and Policy Officer with Tasmanian Parks and Wildlife Service, in six key moments as she develops, runs and reviews a Staff Ride for the first time for the leaders in the land agency's fire management section. This provides some detail of the Staff Ride, highlights some of the issues in running a Staff Ride for the first time and the layers of reflection needed to understand the deeper reasons behind why decisions are made during an incident.

Moment 1: Designing a Staff Ride

Sandy was asked to organise a Staff Ride by her boss who had been on one in the USA and believed that it would be a useful learning tool. After reading up about Staff Rides in the USA and watching videos of some of the Staff Rides, she considered what incident might be suitable and worked with Burn Boss, Phil, to develop a narrative of the Narawntapu fuel reduction burn which had burnt outside the intended burn area, putting a house at risk. While she would have liked to have included other people's voices in telling the story, their availability was a problem and Phil ended up being the sole presenter. The Staff Ride was organised to complement the Parks and Wildlife Service Fire Management's planned burning debrief annual review of their managed burn programme for the year, which was held at the Narawntapu National Park Visitor Centre.

Sandy said 'Our typical processes for reviewing what we do is sitting down with lots of paper, or listening at long meetings. It is not the best environment to learn in and doesn't necessarily suit our different learning styles.' Sandy thus saw the Staff Ride as an opportunity to review operations in a more lively way that could create deeper learning. It tied in with the Department's aspiration to move into being a High Reliability Organisation (HRO) (Weick and Sutcliffe 2001); helping to foster a more reflective culture, where people are willing to consider past mistakes, to see failure not as blame, but rather as weak signals of deeper issues. Thus people were introduced to HRO on the afternoon before the field trip after a day of debriefing their previous season's fuel reduction burns.

The Narawntapu Staff Ride followed the three stage format, preceding field trips and integration sessions. The night before the field ride participants were given preliminary reading of the two-page burn plan and the season debriefing before heading out to the field trip the following morning. This was followed by a short integration phase on site. Although an after-action review had been done on the incident it was decided not to share this with participants in order to allow the story to unfold so that they would not be bringing hindsight bias.

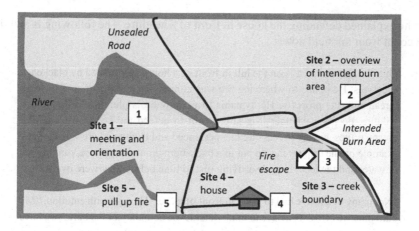

Figure 10.1   Some of the Narawntapu Staff Ride stands

Moment 2: The Field Trip – Generating Reflective Inquiry

Once out at Narawntapu on the morning of the field trip, Sandy set the scene, explaining the rules of conversation – no blame or criticism, rather aiming to understand – and handed over to Phil to orchestrate the story telling. Phil first took people to a vantage point of the planned burn area and then moved to key points where the situation unfolded. At each location Phil described the conditions, what was happening at that moment and what he was thinking, deciding and doing. He invited and posed questions and participants were able to discuss. Everyone was mindful of the way they asked questions. Between the stands there was talk amongst small groups as they walked or drove – the walking was important in getting a feel for the land and the vegetation. In the early stands participants' questions were around trying to make sense of the context, seeking out more details and sharing information about similar incidents.

At Stand 3, at the site of the creek burn boundary where the fire had appeared on the wrong side after the helicopter had dropped its incendiaries to burn the edge, participants became investigators. They looked at the ground around the creek, determining that the incendiaries had been dropped on the wrong side of the burn area edge and then discussed what might have gone wrong in the helicopter, for example a navigation issue, or a technical failure of the incendiary dispensing device. The conversation became very detailed.

Phil explained that he did not know why the fire was on the wrong side of the boundary at the time. His only information, which was radioed from his crew (who were at the house on the hill, outside the planned burn area) was they could see a fire front coming towards them. Phil explained how he decided to head to the house to check out the situation. At Stand 4 Phil's account became quite dramatic as he explained defending the house in front of a headfire. The following is a vivid account from my field notes:

> Phil, the Burn officer, is on the hill in front of a house surrounded by blackened eucalypts. He points to where the fire was coming from at the time of incident threatening the property. He explains to us how he is phoning in that it is a wildfire, deciding whether it is safe or not to defend the house, working out escape routes, having to reassure inexperienced and frightened crew, leading by example on the drip torches to put in a back-burn around the house, running out of water and the local brigade arriving just in time before they were over-run.

> "Not many of us have had to stand in front of a fire." He says with emotion, "My burn plan was crap. You have to think outside the boundary of the burn, you have to consider worst case."

Following this dramatic account the group were asked 'What are you thinking?' The quality of inquiry and conversation shifted out of the detail into considering the larger implications. Topics that emerged included the problems when planned burns switch to wildfire, leadership, successional plans and resourcing.

Moment 3: Unpacking the Processes of Learning

Sandy was surprised afterwards at what came out of the field trip. It revealed details and evoked discussions and insights beyond what she and Phil thought the key message might be (such as, when planning a burn it is critical to think outside the boundaries of the burn area) and beyond what the after-action review had previously revealed. Sandy then wondered what she could learn from the processes of reflective learning that occurred in the Staff Ride to bring to other organisational learning situations such as after-action reviews, Lessons Learnt videos and discussions.

The Bushfire Cooperative Research Centre supported a process to extract learning from the Staff Ride through the creation of a Lessons Learnt video and a guide to designing a Staff Ride. This provided an opportunity for further conversations between Sandy and myself where we began to inquire more into the mechanics and intentions of the Staff Ride, asking what we valued and how it might be improved. Particular discussion topics had emerged from each stand on the day and we wondered whether it would have been useful to have pre-planned specific questions or topics at each stand or to allow for natural emergence of topics. Although one voice (Phil's) provided enough detail to help participants reflect on how they might be acting within a similar situation, it would have been better to hear other key voices to get other perspectives in keeping with some other Staff Rides that have been organised in the USA.

We also realised that the 30-minute integration phase at the end of the field trip was not that effective. People were saturated and participants needed to stay another night to enable sleep time prior to a morning session for further reflection and analysis. Many issues that were raised during the discussion could have been better captured for further discussion and analysis. We realised that it was important to use direct facilitation to encourage discussion and new ways of thinking and that there were some generic questions that might be effective in doing so, such as 'What are you thinking?'

Moment 4: Watching the Video – Exposing Deeper Assumptions

An edited video of the Staff Ride was made and Sandy watched it with one of her regional managers, Eddie, some months after the Staff Ride. Though difficult to reconstruct, maps had been developed to show the location of the fire front, including disposition of resources at different times during the incident. With these and the opportunity to sit back and see the whole within a shorter time span than the actual field trip, both noticed something significant. The decision that Phil had made to defend the house was highly risky. Sandy said:

> Until I looked at the video it just hadn't occurred to me that this was an entrapment event and a safety issue. Phil and all of us on the Staff Ride seemed to have lost sight of the fact that the crew were trapped when they tried to defend the house. Losing a house would have been bad, losing a crew would have been tragic. Although they did defend the house and no-one was hurt, the decision to defend the house should have not been made. The timely arrival of a brigade with water was critical to that success.

Eddie said:

> At the Staff Ride I thought I would make the same decisions as Phil. You just do it one step at the time and each is a logical extension of the next. I think you would have to have been thinking very differently to make any other decision.

So what was it that enabled this 'assumption' to be named up? Did the very construct of the Staff Ride, in discouraging criticism of decisions, preclude the ability of the Staff Ride participants to stand back and ask what assumptions are we making here? Phil's decision in the heat of the moment could be explained as typical of cognitive biases such as being captured by the present, tunnel vision, following a particular train of thinking or part of a deeper, entrenched set of tacit organisational culture – 'we have to protect property at all cost because last time something burnt down it created a major political and community outcry' (see Frye and Wearing, Chapter 4, this volume for a discussion of biases).

Part of the purpose of the Staff Ride is to help people walk in the shoes of those who were at the incident and understand why they may be doing things. But this example shows how it might take time and different kinds of artefacts from experience before it is possible to see beyond the horizon of one's thinking.

Different theories, such as cognitive biases and their impact on decision making, provide the language to explain what is happening without blaming people or their decisions, thus enabling deeper exploration into previously unquestioned culture or policy. This also shows how reflection is layered; it is important to have time, opportunity and new ways of seeing to drill down into or to 'mine' these different layers.

Based on this insight we realised that there might be at least two layers of questions suitable for the Staff Ride (see Figure 10.2 and Figure 10.3). The first engages an open reflective orientation for the first part of the field trip, and the second layer, suitable for the integration phase, engages a more critical orientation where participants are more aware of the frames that they might be bringing and can critically reflect on these. Examples of questions that might be asked are mapped on a framework for dialogical inquiry that explicates different ways of inquiring (Stack and Bound 2012).

Moment 5: The Next Burn Season

Sandy believed that the Staff Ride, despite its simplicity, had considerable impact on many of the participants, with Phil's core message taken on board. Three burns escaped the following year, however, in all of them the consequences were reduced because the risks outside the boundary had been foreseen and mitigated, indicating a shift in perspective (and practice) by the participants. Further, long after the Staff Ride, participants would recall or question something from the Staff Ride, ask for

Figure 10.2   Questions participants could be considering on the Staff Ride field trip

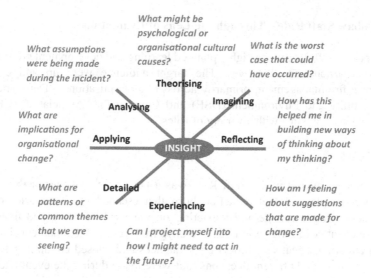

Figure 10.3   Questions to encourage integration

another one to be organised, or suggest an incident that might be suitable. It was a memorable shared experience that people could refer to.

Moment 6: Developing New Processes

Sandy has continued to experiment with creating reflective occasions for her staff drawing on the above lessons. She has developed a process for reflective onsite after-action reviews for those involved in an incident, ensuring that the participants have a night between the field trip and the next day integration session, which involves deeper analysis and pulling out recommendations for organisational learning (Whight 2012). The willingness of people to be involved and the sort of discussion now possible is part of a five year long cultural shift in the Fire Management Department to move towards High Reliability Organisation principles. It is also a testimony to the trust that her staff places in her to treat potentially thorny situations with sensitivity, drawing out productive learning rather than blame.

Sandy cites her exposure to the *human factors* research through workshops with Mary Omodei and Christine Owen and the dialogue with me (from the field of education) as important in helping her to reframe and develop new language in which to think about organisational learning. This has enabled her to subsequently access literature she would not have previously considered important. Shulman (1992) suggested that the architects of case studies gain considerable learning from the process and this is enhanced when the architects engage in reflective cycles.

The Cobaw Staff Ride – Through the Lens of Evaluation

The Cobaw Staff Ride was a highly planned learning and development programme with well-articulated objectives. The target audience were up and coming leaders in fire management, primarily across two organisations – Department of Sustainability and Environment (DSE) and Country Fire Association (CFA) – inviting a vertical slice with a variety of roles.

The Incident

The incident covered in the Staff Ride was a Department of Sustainability and Environment planned fuel reduction burn that escaped the boundaries of the burn, destroying a cottage and impacting on vineyards, fencing and livestock. The fire event covered a period of three days from ignition and had considerable media attention, negative political consequences and caused continuing friction in the community. Different decisions that were made during the event had their genesis much earlier and were influenced by the socio-political climate, as well organisational doctrine and agency operational factors. It offered a rich case study that could reveal operational level information as well as deeper organisational reflection. The learning objectives for the Cobaw Staff Ride were as follows:

- the staff ride was part of a suite of leadership learning experiences, with the intent that, through providing a rich context for decision making, participants could integrate and embed their learning from leadership, human factors and technical knowledge training courses;
- it could act in lieu of real events in helping participants to build up a bank of 'mental slides' to be used in actual events through Recognition Primed Decision Making (Phillips et al. 2004);
- it would foster a reflective non-blame culture – one of willingness to examine past mistakes;
- it would build knowledge in fire behaviour, human factors and decision making, providing insights that could change practice;
- it would help develop meta-cognition;
- it would encourage greater understanding and inter-operability between agencies.

The Format

The Cobaw Staff Ride followed the three distinct phases of Staff Ride learning programmes, but had its own unique approach:

Phase 1: Preliminary study – reading and orientation
A 120-page booklet with information about the incident was sent out to participants two weeks before the field trip. Participants were asked to consider what they

thought might be key decision making points, what stood out for them and what questions they had, as well as to complete a pre-Staff Ride survey. The booklet included information about Staff Rides, human error and biases, weather, climate, geography, the timeline of the incident (based on interviewing people involved in the incident), the after-action review, media reports and pictures at different stages.

The afternoon before the field trip an orientation was given to set up the 'rules' for the field trip discussion, emphasising the need move out of 'hindsight bias' and judgmental thinking into a more open and listening orientation to the story and to each other. The metaphor of 'sliding doors' was used to encourage people to imagine alternative decisions and their consequences. This was followed by an introduction to particular theories (human error and biases, decision making, 'mental slides', recognition primed decision making), providing some meta-cognitive language to enable discussion about one's own thinking. Participants then broke into four groups with a facilitator to get to know each other, and to develop an overview of the unfolding of the incident and the critical decision making points, drawing on their reading.

Phase 2: Field trip (the following day)
Each group headed into the Cobaw National Forest via mini-bus, and walked/drove to different locations (stands) in sequence of the incident. Each stand was chosen to represent an identified decision making point and a particular operational role. The facilitator invited their group to imagine themselves in a particular role (e.g., Crew Officer, Burn Officer in Charge, Regional Fire Manager), gave information pertinent to that role and time to establish a context, and invited the group to discuss and imagine what they might do. The discussion included role plays to test out how conversations might have gone between different people in the scenario. At the last stand an original actor in the incident told their story, which was quite emotive and powerful.

Integration

Only about 30 minutes was given for integration of the event through sharing of insights (which the organisers referred to as 'Cobaw moments') at the end of the field trip. However, participants had a chance to further reflect on their experience through the post-Staff Ride survey. This was considered the weakest aspect of the programme by the organisers, who were juggling trade-offs with budgets and time. One facilitator suggested that there needed to be about 30 minutes between the last stand and the declaration of Cobaw moments to enable personal reflection. He also suggested that participants could be given a pocket book to jot down reflections during the ride.

The Quality of the Reflection – Observations

During the orientation, field trip and integration sessions I moved between different groups, staying with one group for each session, or stand, to hear how the discussion progressed. Throughout the whole programme it appeared to me that participants were keen to engage in the reflective rules of the programme, taking seriously the need to try to let go of hindsight bias in order to try to walk the shoes of the original actors and understand their circumstances and thinking. However, it was initially difficult, particularly in the orientation phase. The pre-survey questions and the booklet had positioned many of the participants into an analytical and criticising frame of mind: *Why wasn't the dozer there? Why wasn't the community engaged? Why wasn't the fire attended to overnight?* Many on the first afternoon were still trying to make sense of the complexity of the incident, wanting to understand the sequence, the decisions and consequences. However, in trying to work out an objective overarching view of what happened it was easy to slip into an evaluation that appeared quite judgemental of actions. I heard participants say something and then correct themselves, saying things such as 'Oh, that is hindsight bias'. I felt that it was important to have this language at this juncture for people to come back to and for others to hear. It also provided an opportunity for practicing meta-cognition – awareness of one's own thinking.

I felt that during the field trip that there was a discernible shift in the quality of reflection – more thoughtful, teasing out nuances, unravelling deeper issues. The following is an example of my field notes:

> We have just walked deeper into the Cobaw forest along a rutted 4WD track past tall messmate eucalypts with a gully below. It feels enclosed within the hills with limited visibility. The guy next to me says 'Look at the vertical fuel load in these trees – that bark is like a wick.' The trunks are blackened, some all the way to the crown. We are at the position where regrowth forest, which had intended to be excluded from the plan burn, had caught fire as a result of spot fires leaping 10 to 20m from the crown of trees.

> We are asked by the facilitator to consider ourselves in the shoes of the crew leader arriving that morning. He gives us information to set the scene. We have just put in a containment line around the regrowth forest, and need to address the problems of continued spotting setting off new fires coming up slope, much of which is not visible through the smoke. The dozer is in danger of being burnt over. 'What are you dealing with? What are you thinking? Why are you thinking that?' the facilitator asks.

> As participants in the group start to tease out what they might know and possible strategies, someone says 'You are not in a position here to see the bigger picture – you are being captured by the immediate present. Whose responsibility is it to

help pull people out of that – the people on the ground here, the Burn Officer in charge or the area manager?'

The environment was a stage, closing down the objective overarching view to different perspectives, so it was easier for participants to get a sense of the situational awareness of the people in the incident, and thus try to understand their thinking. Participants were able to discuss decision making options that they were considering and why, and others could add their perspectives, sometimes in juxtaposition enabling reframing to occur, and giving opportunity for people to change their minds. A participant might suggest a particular strategy and a role play would be conducted to test it – for example, the Burn Boss onsite communicating via radio to the Regional Manager offsite. However, this revealed the normative nature of radio talk where it was difficult to raise concerns or share situation awareness.

During the integration phase, when participants shared their insights or 'Cobaw moments', many demonstrated a self-aware reflective quality:

- 'How important to pick up the subtle signs and interpret them'
- 'Being able to pull back and analyse and check in with another person'
- 'What struck me was the importance and the responsibility of the decision points, each like a "sliding door" moment, each of which determined the whole future of the event (e.g., the decision to light)'
- 'Question what you're seeing. Communicate what you're thinking'
- 'I learnt that not everyone knows what I know, so I should not presume this'.

However, there were some missed opportunities for deeper or critical reflection. In particular, discussion at a latter stand flagged some issues and tensions between the two agencies' perspectives in thinking about and relating to the community. Unfortunately, there was not the opportunity to explore the issue further or consider the underpinning assumptions and organisational doctrine and culture.

The initial insights made by the participants on the day were quite raw and personal, but in the post-Staff Ride survey participants began to bring a more analytical and nuanced eye to their experience and to the implications for them, indicating the value of encouraging reflection after some time had elapsed. For example, one participant wrote:

When planning a fire, I have not paid that much attention to the weather or resourcing. I have usually left that up to the Burn Officer In Charge (OIC). Since the Staff Ride I see how much reliance we put on the Burn OIC's capabilities and capacity to carry out the plan safely. I now see their role from a different perspective – I wasn't aware of the pressures they were under. In future I will endeavour to provide as much information as possible to assist them with their decision making, ensuring the Burn OIC is comfortable with the plan and have all the information they need. I will ensure they are resourced properly and encourage our Operations personnel to take more of an active role in the plan's delivery.

Evaluation of the Programme

Following are summaries of some of the key points from the evaluation of the pre and post-Staff Ride surveys (see Stack and Owen 2012 for more details) to give a sense of how the participants perceived the programme.

Participant reactions

The programme was highly valued by participants with 98 per cent (n = 64) indicating they would recommend it to others. For many the opportunity to hear perspectives and decision making processes of participants in different roles, levels and organisations was highly valuable:

> I valued liaising with senior staff of different agencies, discussing strategy with experts and not so expert exponents in an environment tailored to generate constructive analysis without attributing blame.

Participant learning and impact

Participant learning was individual; people took away different understandings or insights, with many outlining their intentions to change their practice or thinking at a personal and/or leadership level. Others indicated that they had made changes to their practice prior to the Staff Ride.

A Wilcoxin Signed Rank Test comparing the second Staff Ride to other recent learning experiences showed significant positive difference in the degree of perceived personal impact, causing reflection, challenging assumptions and motivating to change. Figure 10.4 compares the mean rating for the Staff Ride with the participants' most recent learning experience for those items that were significant.

Shift in thinking about the incident

Over half the participants changed their views about the incident – from seeing mistakes and assigning blame in the pre-reading stage to acknowledging the complexity of factors and the role of human error in decision making. Many commented that they could imagine themselves in the same shoes making similar decisions. For many there was a change in their framing and many indicated a greater meta-cognitive awareness:

> My Cobaw moment was realising that I have been experiencing some of the human factors previously but not realising it. I'm sure this kind of event could have happened at one of my burns. When reading the pre-course material the decisions seemed so obviously wrong, but when walked through the event, the decisions seemed quite reasonable based on the perspectives of those

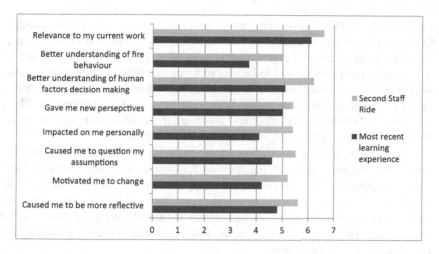

**Figure 10.4   Comparison of the second Cobaw Staff Ride with participants'
most recent learning experience**

involved. This has generated a greater awareness of human factors and how
these influence my decision making. I will undertake a more conscious review
process, particularly when I am involved in high risk decision making.

Developing cultures of reflection

In the pre-survey 40 per cent of respondents indicated that their organisations
buried what happened when things went wrong, and 28 per cent said that blame
was assigned to people. However, the Staff Ride was seen by the participants to
create a positive model for reflective inquiry that they wished to bring to their own
areas of influence. For example, 88 per cent of respondents indicated that they
would be interested in using their own incidents to be part of Staff Rides and over
half were interested in facilitator training:

> In 10 years as an instructor I have rarely seen the impact on people as this type
> of learning – their change in thinking, their willingness to reflect and inquire
> and their curiosity to learn more. It is highly effective and efficient, compared to
> longer courses. (Facilitator)

However, it was too early to see in the main what transformative impact this might
have on behaviours and organisational performance.

Role of the Facilitator

Participants indicated that facilitator knowledge and skills were critical in ensuring
an effective learning experience. The format put considerable pressure on the

facilitators to be fully conversant with the incident; to understand theories such as fire behaviour and human decision making and to know how to bring these in; to know the different fire and organisational roles of people in their group to be able to juxtapose perspectives; and to be able to listen carefully to know when people need more time to make sense of information and when it was possible to move into a more critical reflection phase. With each running of the programme (three times in 2012) this became easier and feedback from participants and observers were important in highlighting areas that could be improved.

Peter, the organiser and designer of the materials, in hindsight said that the pre-reading resource was too big, putting an onerous demand on participants and the facilitators to make sense of a complex situation. In doing it again he would use narratives of people to counterpoint factual timelines of events and decisions to create more emotional rather than just analytical engagement with the case upfront. For more discussion on the role of emotion in decision making see Douglas, Chapter 5, this volume.

Key learning dimensions of the Cobaw Staff Ride
Four key learning dimensions were identified from the participants' experiences. These are summarised in Figure 10.5.

Figure 10.5 provides a simple model of how a case study can seed personal and organisational change through encouraging new insights, modelling and providing practice in meta-cognition, modelling and providing a non-blame inquiring culture and providing rich case studies that can be used as 'mental slides' in time-poor situations.

Implications for Practitioners and Instructors

Staff Rides are examples of learning programmes that are not structured around developing pre-determined understandings. Rather, they provide rich case studies with diverse participants in dialogue so that emergent understandings can unfold. These may be shared by the group but are often quite personal. The role of the facilitator is to orchestrate the conditions for discovery, maximising the potential for people to extend their frames and to hold conflicting views in dialogue, leading to insights that have transformative capacity.

So what are important aspects of facilitation that need to be taken into account for this type of learning? Some aspects include:

1. Set up the rules of engagement for learning environments that foster reflection, non-blame and permission to change one's mind.
2. Establish a shared 'meta' language that enables talking about the processes that we are using. (e.g., I am catching myself using hindsight bias)
3. Be aware that the type of activities, styles of learning or questions that people might experience when engaged in reflective inquiry can shape and

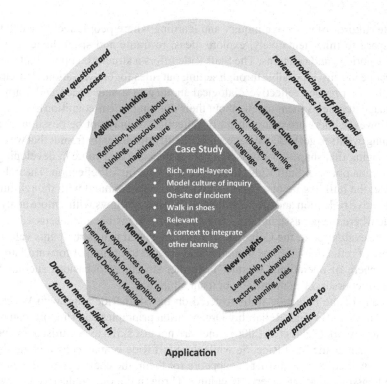

Figure 10.5 Learning outcomes of the Staff Ride and their application

structure the nature of dialogue, and in particular open up or close down conversations or thinking.

4. Develop one's capacity as a facilitator to listen in ways that recognise different phases of reflecting (e.g., making sense, investigative inquiry, imagining, re-framing, assumption hunting) and develop capacity for timely intervention, such as encouraging imagining of alternative decisions and their consequences, or juxtaposing different perspectives to help people loosen their reading of situations to develop a more dialogical stance.

5. Engage in reflective processes about one's facilitation, drawing on models of learning, re-examining one's own orientations to learning and dialogical processes, widening one's view of the type of reflection possible (see Henderson and Kesson 2000).

Conclusion

The challenges of working in an action-oriented and sometimes reactive industry, subject to investigative inquiries and legal review, is that it can be difficult to create cultures of reflective inquiry and learning where people feel safe and have the space to think tentatively, explore ideas, re-frame and speak honestly about their emotions and concerns. The Staff Ride offers a structured opportunity where reflective inquiry is possible through setting out rules for conversation. Participants are able to practice reflective dialogical inquiry in a safe space and learn new language that helps them to talk about their thinking (meta-cognition).

However, there are many layers of reflection. It may not be possible in short learning events to get to deeper layers and identify the different frames that we might be bringing that shape what we see. There is considerable value in developing the integration phase of Staff Rides to provide this layering of reflection. The field trip enables the building up of a trusting, non-blame environment with thoughtful and explorative reflection and then the Integration phase follows with a more analytical frame where deeper assumptions can be made more visible and discussed.

The evaluation of the Cobaw Staff Ride suggested that participants valued the culture of reflection that was modelled and wished to bring that forward into their own sphere of operations, whether in further structured occasions such as Staff Rides or courses, or in mentoring or team situations.

Many of the reflective skills fostered on the Staff Ride align with Weick and Sutcliffe's (2001) High Reliability Organisation principles and support features of good teamwork (Salas et al. 2006) – they are not just skills to be utilised for special events such as the Staff Ride. So while the Staff Ride is an excellent beginning to model such cultures and whet the appetite for them, the challenge for the industry is to transform that into everyday culture. Through ongoing reflective dialogue it is possible to uncover frames, consider new ones, test them out and through the process grow individual and organisational capacities.

Acknowledgements

The research was supported with funding from the Bushfire Cooperative Research Centre. However, the views expressed are those of the author and do not necessarily reflect the views of the Board of the funding agency.

References

Argyris, M. and Schön, D. (1974). *Theory in practice. Increasing professional effectiveness.* San Francisco: Jossey-Bass.

Brookfield, S.D. (1995). *Becoming a critically reflective teacher* (1st edition). San Francisco: Jossey-Bass.

Boud, D., Cressey, P. and Docherty, P. (eds) (2006). *Productive Reflection at Work.* New York: Routledge

Boud, D. (2006). Creating the space for reflection at work. In D. Boud, P. Cressey and P. Docherty (2006). *Productive Reflection at Work.* New York: Routledge, pp. 158–169.

Caffarella, R.S. (2002). *Planning programs for adult learners.* San Francisco: Jossey-Bass.

Dekker, S. (2006). *The field guide to understanding human error* (2nd edition). Aldershot: Ashgate Publishing Limited.

Dewey, J. (1933). *How we think.* New York: Heath.

Eraut, M. (2000). Non-formal learning and tacit knowledge in professional work. *British Journal of Educational Psychology* (2000), 70, pp. 113–136.

Frye, L. and Weaving, A. (2011). The Central Mountain Fire Project: Achieving cognitive control during bushfire decision making. *Cognitive Technology,* 16.

Henderson, J.G. and Kesson, K.R. (2004). *Curriculum wisdom: educational decisions in democratic societies.* New Jersey: Pearson.

Hoyrup, S. and Elkjaer, B. (2006) Reflection:taking it beyond the individual. In D. Boud, P. Cressey, and P. Docherty, *Productive Reflection at Work.* New York: Routledge, pp. 29–42.

Kegan, R. (1994). *In over our heads: The mental demands of modern life.* Cambridge, MA: Harvard University Press.

Kemmis, S. (1985). Action research and the politics of reflection. In D. Boud, R. Keogh and D. Walker, *Reflection: Turning experience into learning.* Oxon: Routledge Falmer, p. 139.

Killen, R. (2003). Chapter 3. Becoming a reflective teacher. In Thomson, *Effective teaching strategies.* Sydney. pp. 48–61.

Kirkpatrick, D.L. (1998). *Evaluating training programs: The four levels* (2nd edition). San Francisco: Berrett-Koehler.

Klein, G., Ross, K.G. et al. (2003). Macrocognition. *IEEE Intelligent Systems,* 18(3), pp. 81–85.

Kolb. D.A. and Fry, R. (1975). Toward an applied theory of experiential learning. In C. Cooper (ed.), *Theories of group process.* London: John Wiley.

Omodei, M. (2012). *Introduction to human factors: Individual.* A learning manual. BCRC

Phillips, J.K., Klein, G. and Seick, W.R. (2004). Expertise in judgement and decision making: A case for training intuitive decision skills. In D.J. Koehler and N. Harvey (eds), *Blackwell handbook of judgement and decision making.* Victoria: Blackwell Publishing, pp. 297–315.

Reason, J. (2008). *The human contribution.* Aldershot: Ashgate.

Salas, E., Rosen, M.A. et al. (2006). The making of a dream team: When expert teams do best. In K. Anders Ericsson, N. Charness, P.J. Feltovich and R.R. Hoffman (eds), *The Cambridge handbook of expertise and expert performance.* New York: Cambridge University Press, pp. 439–453.Scharmer, O. (2009). *Theory U: Leading from the future as it emerges.* San Francisco: Berret-Koehler Publishers.

Schon, D. (1983). *The reflective practitioner: How professionals think in action.* New York: Basic Books.

Senge, P. (2004). *Presence: Human purpose and the field of the future.* Cambridge: Society for Organizational Learning.

Shulman, L.S. (1992). Toward a pedagogy of cases. In J.H. Shulman (ed.) *Case methods in teacher education.* New York: Teachers' College Press, pp. 1–30. Retrieved from http://blogs.shu.edu/marsh/files/Chapter%201.pdf.

Stack, S. and Bound, H. (2012). *Exploring new approaches to professional learning: deepening pedagogical understanding of Singapore CET trainers through meta-cognition and practitioner-based research.* Report for Institute of Adult Learning. Retrieved on 9/5/2012 from: http://www.ial.edu.sg/files/documents/457/TLD%20Report_External%20Report.pdf.

Stack, S. and Owen, C. (2012). *Evaluation report: 2012 Cobaw Staff Ride program.* Bushfire CRC.

Stack, S., Owen, C. et al. (2010). *Designing the Staff Ride: A vehicle for learning from wildfire and prescribed burning operations in Australia.* Bushfire CRC. Retrieved on 12/02/2013 from http://www.bushfirecrc.com/resources/guide-or-fact-sheet/designing-staff-ride.

Sutton, L. and Cook, J. (2003) *Wildland Staff Ride guide.* Retrieved on 27/02/2013 from http://www.wildfirelessons.net/SearchResults.aspx?q=staff%20ride.

Taylor, K., Marienau, C. and Fiddler, M. (2000). *Developing adult learners: Strategies for teachers and trainers.* San Francisco: Jossey-Bass Inc.

Tobert, B. (2004). *Action inquiry: The secret of timely and transformational leadership.* San Francisco: Berrett-Koehler.

Yorks, L. and Marsick, V.J. (2000). Organisational learning and transformation. In J. Mezirow (ed.), *Learning as transformation: Critical perspectives on a theory in progress.* San Francisco: Jossey-Bass Publishers, pp. 253–284.

Weick, K. (2002). Human factors in fire behaviour analysis: reconstructing the dude fire. *Fire Management Today,* 62(4). Retrieved 27/02/2013 from http://www.fs.fed.us/fire/fmt/fmt_pdfs/fmt62–4.pdf.

Weick, K.E. and Sutcliffe, K.M. (2001). *Managing the unexpected: Assuring performance in an age of complexity.* San Francisco: Jossey-Bass.

Wenger, E. and Lave, J. (1991). *Situated learning: Legitimate peripheral participation.* New York: Cambridge University Press.

Whight, S. (2012). *Better learn from this one: Creating a culture of lessons learned in a sceptical workplace.* Retrieved on 12/03/2013 from http://www.bushfirecrc.com/resources/presentation/better-learn-one-creating-culture-lessons-learned-sceptical-workplace.

Chapter 11

The Challenges of Change in Future Emergency Management: Conclusions and Future Developments

Dr Christine Owen
Bushfire Cooperative Research Centre, Australia and University of Tasmania, Australia

Dr Karyn Bosomworth
Bushfire Cooperative Research Centre, Australia and RMIT University, Victoria, Australia

Steven Curnin
Bushfire Cooperative Research Centre, Australia and University of Tasmania, Australia

Introduction

As the chapters in this volume have discussed, the world of emergency services responders and managers is complex. Decisions are frequently made under ambiguous, dynamic and uncertain conditions involving trade-offs, with the consequences impacting many people. As was discussed in the Chapter 1 these complexities are increasingly driven by a number of trends. Those trends include climate change, which exacerbates the vulnerabilities of communities and our environment intersecting with other dynamic influences such as global socio-economic shifts, demographic changes, and biological system shifts (Few 2007). Another trend comes from increased interdependencies of social, technical and infrastructure systems. For example, there is a strong reliance on energy in our society for food production, transport and even social interactions. Interconnectivity of urban systems also means that negative impacts on one system could have a cascading affect and influence on the functioning of other systems. These drivers have implications for the future context of emergency services. This volume contributes practical suggestions to enhance the services' capacity to manage large scale emergency events in this future world as well as pointing to areas in need of further attention.

Challenges for Emergency Management

To conclude the book, we take seven themes evident in the literature as future trends impacting on emergency services and consider what the contributions here offer for managing these trends in the future.

Increased Complexity and Uncertainty

As discussed in Chapter 1, the number and intensity of adverse events is increasing, that is, extreme weather events or complex technological and infrastructure breakdowns. The report 'Hardening Australia' (Yates and Bergin 2009) noted that disasters are likely to become larger, more complex, occur simultaneously and impact on regions that have either not experienced natural hazards previously or at the same intensity or frequency. Moreover, social and ecological vulnerabilities to those events are also increasing.

These changes make the work of emergency services personnel more challenging. Bosomworth and Handmer (2008) argued that climate change will not only likely increase the risks of natural hazards such as bushfires and their associated impacts and losses, it will also likely impact underlying vulnerabilities and resilience. For example, combined with globalised economic pressures and drought losses, a natural hazard may represent a threshold beyond which an individual or community cannot cope or recover. Climatic driven changes in land use may contribute to rural decline, further reducing social networks and volunteer numbers. Essential services that contribute to a community's resilience face increased threats, and already stressed ecosystem functions and services may decline further, with concomitant impacts on social resilience.

Increased Vulnerabilities

These future trends are likely to put greater pressure on emergency management teams particularly as demands on emergency services increase. Obtaining resources for managing future emergency events will be a challenge in the future in four respects:

- Workforce rationalisation and economic cutbacks have impacted on the government and agencies who have historically supplied such personnel.
- The adversarial nature of several post-event inquiries have taken their toll on the motivation levels of current staff to put themselves into positions of decision making authority.
- Organisational restructuring and downsizing has reduced outsourcing and supply of traditional services and resources that are within the organisation.
- Demographic changes affecting communities also impact the potential resource base of emergency services organisations. In many countries, and especially in Australia, the emergency services sector relies heavily on

volunteers and non-government organisations for prevention, preparedness, response, and recovery. Demographic changes will also mean younger and less experienced personnel will need to step up and manage emergency events sooner than was typical in career progression pathways of the past (Howard 2009).

These trends are going to place higher stressors and cognitive demands on crews, emergency management teams and leaders. Strategies to manage stress and to regulate thinking to avoid biases will become important, particularly as events are of longer duration. McLennan et al. (Chapter 2) suggest elements for inclusion in training. These include using positive self-talk, mental re-evaluation of the situation and assumptions, information or advice seeking, physical movement, calming breathing and effortful-ignoring of issues which are not immediately relevant to the most serious emerging threats.

These trends will also have impacts on the decision making capacities of the people involved. Chapter 3 (by Johnson) and Chapter 4 (by Frye and Wearing) examined potential sources of bias and identified strategies that can be used to support the thinking processes of responders so that they get the best outcomes possible under challenging conditions. Johnson (Chapter 3) provides a resource for responders and managers facing more demanding conditions in her use of worst case scenario thinking for contingency planning. Worst case scenario thinking involves identifying possible worst case outcomes so that strategies can be enacted to reduce the probability of such a negative outcome. In high-risk work environments, these kinds of strategies are particularly important for supporting reliability and avoiding accidents and even death. As she concludes, the importance of higher-order cognition for all decision makers cannot be overstated; developing metacognitive skills has the potential to improve worst case scenario thinking by minimising various barriers. The ability to critique personal decisions is challenging because it requires strong metacognitive skills that allow decision makers to reflect on their own thoughts and feelings.

Frye and Wearing (Chapter 4) discussed the role of metacognition as part of self-awareness when working under the time-critical, dangerous and pressured situations of firefighting. They recommend that emergency services agencies focus on two criteria for developing expertise, namely: learning in context, and learning calibration skills (or metacognitive skills). Their strategies assist practitioners to:

1. recognise trigger points for changing goals (e.g., from containment goals to protection goals)
2. manage risks and responsibilities within a chain of command (e.g., delegation and escalation)

There are also suggestions to accelerate learning for newcomers to leadership positions so that they can calibrate their thinking. By this they mean to learn to think like other more experienced decision makers. They suggested that this type of learning needs to be distributed throughout the group to include peers (i.e., buddy as coach), teams (i.e., crew as coach) and supervisors (i.e., leader as coach), and also from mentors and subject matter experts.

Interdependencies and Converging Pressures

The numbers of stakeholders engaged and involved in emergency management is growing and this will continue into the future, resulting in larger teams and increased interdependencies between stakeholders. Larger teams present greater potential for breakdowns in coordination, particularly when coupled with the other pressures described by Bremner et al. (Chapter 8). Bremner et al. note that in contrast to latent vulnerabilities, pressures may be 'strong or weak, subtle or coercive, direct or indirect' (Paletz et al. 2009, 436). They identify that it is likely that responders and managers must collectively deal with several pressures simultaneously. Some of these pressures come from an increasing expectation for the provision of seamless lateral and hierarchical delivery of services and real-time information to a variety of stakeholders; stemming from increases and diversity in information media streams, as well as the growing number of stakeholders. This places increased demands on organisational interoperability; exponentially increases expectations for forecasts, warnings and real-time information; and increases interdependencies in decision making and thus pressures on teams.

Teamwork communication and multi-agency coordination will become more important under these conditions. As Douglas (Chapter 5) and Hayes (Chapter 6) noted, conflicting goals need to be managed and negotiated. Both these chapters provide some valuable suggestions for practitioners to enhance their team's performance and include resources for use in training interventions to improve performance.

Given that teams perform better if members are familiar with one another (as Hayes, Chapter 6 argues) and that future emergency management teams are likely to grow in size and include people who may not have worked together before, what can be done? Hayes proposes that there may be benefits from conducting team-based training and exercises with personnel from different teams. Requiring members to simulate working in unfamiliar teams can create an environment that requires personnel to learn how to interact more readily and adapt to variations in the conduct of their new colleagues. Hayes also suggested some activities that practitioners could use to facilitate ad hoc teams to become familiar with one another. These included brief résumés and short question and answer sessions. Such strategies can enhance the transactive memory of the team (Levine et al. 2005, see Hayes, Chapter 6) helping newcomers learn who has which task-relevant skills and help teams quickly and successfully integrate newcomers through the generation of swift trust (Wildman et al. 2012).

Social Information Technologies, Networking and Emergence

Social information and communication technologies are rapidly proliferating within emergency services organisations in an effort to overcome previously identified problems of coordination associated with the involvement of multiple stakeholders (e.g., Peek and Sutton 2003, Comfort 2005, Comfort and Kapucu 2006). These problems have included:

- lack of suitable communications infrastructure, including a lack of compatibility between data systems and communications technologies
- communication difficulties between coordination centres and the incident ground
- poor integration of different agencies 'response' plans
- poor and varied levels of situational awareness (including projections) among emergency management partner organisations
- lack of timeliness and accuracy in information dissemination.

In addition, people interested in the emergency events but not formally part of traditional command and control arrangements are now active stakeholders by sharing information through social media. In part, the rise in the use of multiple social media sources by community members during emergencies has been driven by demands for more and faster information that may not be forthcoming from emergency services.

With these changes come challenges in how emergency services organisations must successfully operate with the tensions of potential information distortion, greater intelligence gathering requirements and self-organised emergence of community groups. For example, lateral information networks mean that information comes from diverse and sometimes unverified sources and not necessarily from within traditional operational channels. These trends will continue to challenge the traditional cultural identities highlighted by Douglas in Chapter 5. If not managed well the suboptimal management of these traditional cultural identities can result in communication barriers within teams and can introduce biases in thinking and fragmentation of groups. Practitioners might use the ideas contained in Douglas' chapter to expose and confront any negative stereotypes that can impede team performance.

The proliferation of social information and communication technologies indicates a need for greater attention to external liaison and technologies that support distributed situation awareness in order to support multi-agency coordination specifically involving stakeholders from non-emergency organisations and the general public. The work of Owen in Chapter 7 is of particular significance and responders and managers could use the resources in that chapter to recognise the key boundary spanning and boundary crossing activities of incident controllers and other team leaders in order to build these into leadership programmes including feedback in exercising.

Of particular importance will be strategies to enhance coordination between layers of teams engaged in emergency management arrangements. Bremner et al. (Chapter 8) point to areas where there can be structural and cultural barriers leading to breakdowns in coordination. Practitioners could use the ideas contained within that chapter to examine and address potential areas for coordination breakdown.

Evaluating Emergency Management Response Effectiveness

Given the scrutiny of emergency management processes in post-event inquiries, it is important that those engaged in all levels of emergency management, including political leaders, have well-established and understood processes. Also crucial are agreed process and outcome measures for evaluation of emergency management performance. Absence of these processes and outcome measures represents considerable risk of and exposure to unfair criticism and litigation for emergency services organisations.

As Bremner et al. (Chapter 8) suggest, this is particularly challenging in existing emergency services organisations, with both process and outcomes coming under increasing attack in coronial and other post-event inquiries. Bremner's chapter highlighted the important of developing innovative classification systems to examine the demands and pressures on people in the emergency management environment. His work suggests that establishing criteria to evaluate the *process* and outcomes of emergency events needs attention, alongside the question of who conducts such evaluations. This is often evident in emergency services where practitioners reported that they feel their performance is arbitrarily judged by external sources such as the media, even after the emergency event (Owen et al. 2013).

In many other safety-critical industries, sole reliance on outcome measures has been found flawed and even dangerous to the longer-term viability of the safety-critical system (Dekker 2006, Hollnagel, Woods and Levensen 2006). The same can be said of emergency management. The outcome from an emergency event might have been successful despite risks and unsafe practices being undertaken (and thus luck that there was not an adverse event). Conversely, all the best measures and processes might have been in place and performed well but the outcome might have still had negative impacts because of the nature of the event or the pre-existing vulnerabilities that made avoidance of impact virtually impossible. It is important that those working in senior leadership positions have process and outcome measures to be able to assess whether or not their management objectives are on track.

Coping Ugly under Degraded Conditions

It is vital to acknowledge that during major or extreme events, emergency management operations are frequently degraded (e.g., communications failures;

insufficient resources; escalating and uncontrollable conditions). It is also crucial therefore to acknowledge that in these major events, improved safety occurs through recognition and pro-active management in suboptimal conditions (Brooks, Chapter 9). Within the emergency management system Brooks refers to 'degraded' as a lower level of quality of operations than might be desirable or even necessary to maintain a desirable level of safety. This degradation may be in the information available or received, in the reactions of people to external stimuli, in the assets to manage the event, or other aspects of the coordination system. Brooks also observes that many of the usual supports that help people to do their work are not functioning effectively. Emergency services workers need to make do, even if they are fatigued or technologies are not working properly, or the needed resources are not available. Under these circumstances a new framework for operation is needed and Brooks calls this entering the 'zone of coping ugly'. He concludes the chapter by providing some strategies practitioners might use in training specifically aimed at coping under extreme conditions.

It is also important to also acknowledge that mistakes will be made; that no complex operation can ever be perfect. Personnel need support in managing despite these conditions and in recognising shifts toward unsafe conditions. What is required is adaptive behaviour, effective teamwork coordination, learning as part of the process of adaptive coordination and flexible strategies.

Another resource available to practitioners is contributed by Johnson (Chapter 3) who makes the point that worst case scenario thinking can be useful for such contingency planning and this can be trained for and used in real-time performance. Johnson also points out that application of worst case scenario thinking may not only apply to those in the fire and emergency services sector but also for the people who are affected by emergency events in order to enhance community resilience.

Community Expectations and Resilience

Communities and individuals vary in their capacity to prepare for, respond to, and recover from the impact of hazards. Understanding factors that contribute to these variations and using these insights to support communities in building their resilience is an important objective of the emergency management sector. It is also worth noting that while 'community' is often identified by location in emergency management planning, communities can be defined by forms of identity other than place, such as interests, gender, age and workplace. These non-place-based communities often rely upon disaster sensitive information and communication technologies to communicate during events.

Climate change adds an extra dimension to the challenges for resilience: the capacity to deal with expanding and changing risks, and to change or transform the systems and approaches on which our current situation is based (ones that assume no change, or change within previously experience parameters).

As discussed earlier, climate change is likely to expose communities to severe weather events they have not hitherto experienced. The potential for impacts from unfamiliar hazards on underprepared communities is a significant challenge facing disaster managers. Given that worst case scenario thinking can be challenging for experienced incident managers, it is likely that community members may also find anticipating extreme events problematic. In related research reported in Chapter 3, Johnson reports on interviews with community members involved in the Black Saturday 2009 bushfire disaster in Victoria, Australia. Johnson notes that it was clear from some of the interviews that prior to the fires, many people were unable to imagine how the situation could quickly deteriorate into disaster. Many survivors reported difficulties in planning for extreme outcomes and developing back-up options. Therefore, an important application of these concepts for practitioners may be to draw on worst case scenarios to support community decision-making. Further research could investigate factors that might encourage community members to prepare for worst case scenario and provide guidance for developing new ways to improve worst case scenario thinking. Such work would build on scenario planning methods developed in the climate change adaptation arena (see for example Wiseman et al. 2011).

Development and Capability

All of these trends present challenges and place new demands on the sector's leadership and capability, and set up challenges for leadership and capability development. This includes a need to explicitly shift from the traditional focus on event management to a more political, strategic planning and policy focus.

A number of reports suggest that tactical and operational training is well established, particularly for routine events (Murphy and Dunn 2012). However, the same cannot be said for novel or extreme events. A lack of training for such events, at all levels, places undue stress on people who care deeply about the outcomes.

Brooks (Chapter 9) makes the point that emergency response and its management faces challenges that are in some respects, similar to those in other safety-critical domains and in some respects unique. The non-technical skills such as leadership and communication are necessary irrespective of the domain. However, the complexity of the work environment, the distributed nature of the teams and the associated decision making, and the naturally eroded safety management system are just three examples of why approaches adopted in other domains cannot simply be applied within emergency management. It is important to understand what future trends mean regarding the new skills required to successfully operate in this environment. These include the requirement to provide expertise in multi-jurisdictional and cross-agency strategic emergency management arrangements. Some of the findings reported here provide some valuable future directions.

For firefighter safety training, survival mode procedures need to be simple, and be practised frequently under realistically simulated life-threat conditions (McLennan et al., Chapter 2). Training needs to support responders to become skilled at recognising indications that they are experiencing increased stress levels and may need to take effective emotional self-regulation actions (McLennan et al., Chapter 2). In addition McLennan et al. (Chapter 2) also conclude that possible stress-related decrements in judgement and decision making quality need to be taken into account in any after-action debriefs and reviews, and post-incident investigations.

For operations, worst case scenario thinking (Johnson, Chapter 3) needs to be cultivated in order to foster anticipation of rapid escalations of threat situations. Johnson (Chapter 3) suggests a range of procedures to embed worst case scenario thinking into agency practice. Her chapter includes tangible examples of what fire and emergency services agencies can do to enhance resilience.

The leadership and capability needs for emergency management teams require skill, capacity and leadership development; including the ability for personnel to recognise shifts towards degraded conditions and requirements for collective recovery.

Brooks' (Chapter 9) suggestion to re-examine existing training pathways will be particularly useful for practitioners to consider. In addition, training to enhance thinking and to improve team communication and team performance can include the suggestions by Hayes (Chapter 6) and Owen (Chapter 7) to support team communication, coordination and decision making. This requires greater attention to developing skills in complex problem solving, in leadership and communication. In addition, addressing the emotional challenges faced by emergency services personnel when they are dealing with large and rapidly escalating events will be increasingly important. Developing strategies to raise awareness about the impact of emotion on individual and team decision making as suggested by Douglas (Chapter 5) will be advantageous for emergency management practitioners.

The chapters here also emphasise the value of reviewing past bushfire incidents. Every prior emergency incident provides a substantial resource for organisational learning. However, this is not easy in agencies where organisational cultures are typically reactive. This is driven in part because emergency events are characteristically limited with opportunities (time) to stop, critically reflect and adjust. In Chapter 10 Stack discusses some resources to assist practitioners in reframing, as well as enhancing and enhancing reflective learning to facilitate change.

Future Directions for Research and Practice

The demands associated with incident complexity, managing uncertainty, community and political expectations, and changes in the sector's cultural identity, present new challenges. These challenges impact leadership and capability development for practitioners and policy developers, as well as leaders, and suggest areas in need of future research.

As McLennan et al. conclude (Chapter 2), it seems highly desirable that researchers increase our knowledge and understanding of the effects of threat-related stress on bushfire safety through studies focused on firefighters' experiences and actions on the fireground. In addition, further effort is required to evaluate the impacts of strategies such as worst case scenario thinking (Johnson, Chapter 3) and metacognition (Frye and Wearing, Chapter 4) to evaluate its efficacy on individual and team performance. The role of emotion in individual and collective performance in safety-critical domains is another area that has had very little research attention (Douglas, Chapter 5). Strategies to improve psychological safety in teams to enhance familiarity (Hayes, Chapter 5) and how team leaders facilitate coordination between teams (Owen, Chapter 6) are likely to be important given future trends.

In terms of team decision making, Bremner et al. note (Chapter 6) it is clear that we currently know very little about decision making at regional/state levels of fire management. It is unclear what decisions are made, how they are made or what pressures may lead to faulty decision making. Given the focus of post-accident and coronial inquiries on decisions made at regional/state levels of emergency management following large-scale emergencies it is vitally important that we rectify this knowledge gap to better support regional/state level decision making.

Further research is needed to understand how strategic levels measure the performance of an emergency management system, and to understand how personnel and the systems that support them may become aware of degradations of performance safety so that coping strategies can be developed (Brooks, Chapter 9). There is also a need to examine longer-term issues with strategic arrangements at a state level and to contextualise emergency management response within broader goals of governance, disaster risk reduction and sustainable development (to address the issues raised in this chapter). Some questions that emerge from this synthesis to support the development of enhanced capability include:

1. Can we teach non-experts to learn like experts? A longitudinal evaluation of training from this perspective could help answer this question.
2. Will tools such as 'coping ugly' (and the associated development of coping repertoires) help improve the effectiveness of emergency management coordination above the Incident Management Team?
3. How close to the edge of chaos is it possible to train while doing no harm to participants?
4. How can performance be measured in unbounded environments where what is avoided is just as important and when outcomes are more often than not, imperfect?
5. What are the experiences of non-traditional and non-prepared members of the community that can help inform the future of emergency management thinking?

Conclusion

As argued throughout this book, the often unpredictable nature of hazards and subsequent impact on emergency services response is not as readily contained as in other safety-critical domains. This tells us something about a potential future world of work that is highly dynamic, interdependent and for which improvisation, critical thinking and problem solving are necessary pre-requisite skills.

Finally, given the indications of increases in extreme weather-related events, the need for a better understanding of human factors in the fire and emergency services domain has never been more timely or needed. In presenting practical suggestions to enhance the services' capacity to manage future trends and challenges for the dynamic domain of emergency services, as well as pointing to areas in need of further attention, this book represents an important contribution to this emerging field of human factors in emergency management.

Acknowledgements

The research was supported with funding from the Bushfire Cooperative Research Centre Extension Grant. However, the views expressed are those of the authors and do not necessarily reflect the views of the Board of the funding agency.

References

Bosomworth, K. and Handmer, J. (2008). Climate change and community bushfire resilience. *Community Bushfire Safety*. Collingwood: CSIRO, pp. 175–183.

Comfort, L.K. (2005). Risk, security, and disaster management. *Annual Review of Political Science*, 8, pp. 335–356.

Comfort, L.K. and Kapucu, N. (2006). Inter-organizational coordination in extreme events: the World Trade Center attacks, September 11, 2001. *Natural Hazards*, 39, pp. 309–327.

Dekker, S. (2006). *The field guide to understanding human error* (2nd edition). Aldershot: Ashgate Publishing Limited.

Few, R. (2007). Health and climatic hazards: Framing social research on vulnerability, response and adaption. *Global Environmental Change,* 17(2), pp. 281–295.

Hollnagel, E., Woods, D. and Leveson, N.G. (2006). *Resilience engineering: Concepts and precepts*. Aldershot: Ashgate.

Howard, B. (2009). Climate change and the volunteer emergency management sector. *National Emergency Response*, Winter, pp. 8–11.

Levine, J.M., Moreland, R. et al. (2005). *Personnel turnover and team performance*. Arlington: United States Army Research Institute for the Behavioral Sciences.

Murphy, P. and Dunn, P. (2012). *Senior leadership in times of crisis*. Deakin West, Australian Capital Territory: Noetic Group Ltd.

Owen, C., Bosomworth, K. et al. (2013). *Strategic level emergency management: Some challenges and issues for the future*. Melbourne: Bushfire Co-operative Research Centre.

Paletz, S.B.F., Bearman, C.R. et al. (2009). Socialising the human factors analysis and classification system: incorporating social psychological phenomena into a human factors error classification system. *Human Factors*, 51(4), pp. 435–445.

Peek, L. and Sutton, J.N. (2003). Disasters: an exploratory comparison of disasters, riots, and terrorist acts. *The Journal of Disaster Studies, Policy and Management*, 27(4), pp. 141–157.

Wildman, J.L., Shuffler, M.L. et al. (2012). Trust development in swift starting action teams. *Group & Organization Management*, 37, pp. 137–170.

Wiseman, J., Jones, R. et al. (2011). *Scenarios for climate adaptation report: Executive summary*. Victorian Centre for Climate Change Adaptation Research, University of Melbourne.

Yates, A., and Bergin, A. (2009). *Hardening Australia: Climate change and national disaster resilience special report* (Vol. 24). ACT, Australia: Australian Strategic Policy Institute Ltd.

Index

Note: page numbers in *italic* type refer to
    Figures; those in **bold** type refer to
    Tables.

out-of-scale events 4–5, 8, 174
overconfidence 38, 39, 103
overlapping-ness **134,** 141
Owen, Christine 1–18, 103, 125–47, 174,
    188, 207, 219–30, 223, 227, 228

Pakistan, floods (2010) 3
Pearce, C.L. 107
perception, in situation awareness 157–8
perceptual-motor skills, and stress 21,
    22–3, 24, 25, **27**
Perry, M. 107
person-based trust 108
    see also trust
Phillips, J.K. 196
physical environment, and decision making
    162
pilots, and metacognition 60, 64, 69
planning fallacy 38
planning, in incident management systems
    9
practical thinking 182–3
pre-mortem method, worst-case scenario
    thinking 48–50
predictability, and teams 105, 107
process loss 105
projection, in situation awareness 157, 158
psychological safety 14, 104–5, 111, 116, 228

Quarantelli, E.L. 1
Queensland, Australia:
    floods (2011) 3
    Tropical Cyclone Yasi (2011) 4
question-and-answer sessions 114, 115, 222

Rail Resource Management (RRM) 189
Rasmussen, J. 179, 185
Reagans, R. 102, 109–10, 112, i113
Reason, J. 172–3, 174, 195
Recognition Primed Decision Making
    (RPDM) 58, 67, 69, 155–6, 159,
    181, 196, 208
reflective learning 15, 195–200
    Staff Ride case studies 200–16, *203,
    206, 207,* **210,** *213, 215*
regional level of emergency management
    8, **8,** 15, 149, 151, 152, 153–5, *154,*
    158–9

and decision making 156, 165–7, 173,
    174, *175,* 178, 185, 228
relationship management 141
resource management 166
responsibilities, in emergency management
    7–8, **8**
résumés 114, 222
risk perception, and bias 37–8, 39
Roberts, K.H. 102–3
roles 7–8, **8**
    and Boundary Riding 134–5
    and Boundary Spanning 137–8
RPDM (Recognition Primed Decision
    Making) 58, 67, 69, 155–6, 159,
    181, 196, 208
rules of thumb *see* heuristics

safety of firefighters 70–1, 75
safety theory 174
Schon, D. 197
Schulman, L.S. 199, 207
Seick, W.R. 196
self-awareness 13, 30, 58, 59, *59,* 75
self-regulation 30, 58, 59, *59,* 60, 64, 73,
    75, 237
    see also calibration
Shappell, S.A. 160, 164, 165
shared (distributed) leadership 107–8
    see also leadership
shooting stars 89, 90, 93
Simon, Herbert 100
Simon, R. 103
Sims Jr., H.P. 107
situation aversion 164, 165
situational awareness 62, 63, *66,* 66–7,
    101, 156, 157–9
Skriver, J. 156
Slovik, P. 184
social environment, and decision making
    162–3
South Australia, Wangary bushfires 36, 48,
    173, 174, *175,* 176
span of control 9, 173
specialisation in emergency services sector
    10
speed-accuracy tradeoff 67–8
Staats, B.R. 102
Stack, Sue 15, 143, 195–218, 227

Printed in the United States
By Bookmasters